The University of Chicago Press CHICAGO AND LONDON

THE PALE BLUE DATA POINT

JON WILLIS

AN EARTH-BASED PERSPECTIVE ON THE SEARCH FOR ALIEN LIFE

The University of Chicago Press, Chicago 60637
The University of Chicago Press, Ltd., London
© 2025 by Jon Willis
All rights reserved. No part of this book may be used or reproduced in any manner whatsoever without written permission, except in the case of brief quotations in critical articles and reviews. For more information, contact the University of Chicago Press, 1427 E. 60th St., Chicago, IL 60637.
Published 2025
Printed in the United States of America

34 33 32 31 30 29 28 27 26 25   1 2 3 4 5

ISBN-13: 978-0-226-82240-2 (cloth)
ISBN-13: 978-0-226-82241-9 (e-book)
DOI: https://doi.org/10.7208/chicago/9780226822419.001.0001

Library of Congress Cataloging-in-Publication Data

Names: Willis, Jon (College teacher), author
Title: The pale blue data point : an Earth-based perspective on the search for alien life / Jon Willis.
Description: Chicago : The University of Chicago Press, 2025. | Includes bibliographical references and index.
Identifiers: LCCN 2025006420 | ISBN 9780226822402 cloth | ISBN 9780226822419 ebook
Subjects: LCSH: Exobiology | Life on other planets | Extraterrestrial anthropology
Classification: LCC QH326 .W56 2025 | DDC 576.8/3—dc23 /eng/20250430
LC record available at https://lccn.loc.gov/2025006420

♾ This paper meets the requirements of ANSI/NISO Z39.48-1992 (Permanence of Paper).

Authorized Representative for EU General Product Safety Regulation (GPSR) queries: **Easy Access System Europe**—Mustamäe tee 50, 10621 Tallinn, Estonia, gpsr.requests@easproject.com
Any other queries: https://press.uchicago.edu/press/contact.html

CONTENTS

Preface [vii]

1 The Pale Blue Data Point [1]

2 Twenty Thousand Pings Under the Sea: In Search of Alien Oceans [36]

3 Swimming with Stromatolites: The Hunt for Martian Fossils [75]

4 The Arc of the Firmament: Mapping Exoplanets [110]

5 To Catch a Falling Star: Meteorites and the Clues to Earth's Origin as a Life-Bearing Planet [153]

6 So Long and Thanks for All the Fish: A Dolphin-Led Guide to Alien Communication [190]

Acknowledgments [225]
Further Travels [227]
Index [233]

Color plates follow page 112.

PREFACE

Having written one book about astrobiology in which no alien life was discovered, why do I now find myself writing another? Do I have any secrets up my sleeve? Any breaking astrobiological news? Well, one can always hope, but for the present the answer is no. However, as I wrote *All These Worlds Are Yours* in 2016 I realized that there was a companion story to be told to accompany the classical account of astrobiology as taught in college classrooms. This story was very much an adventure, one scattered across dramatic landscapes on planet Earth. Our pale blue data point offers many opportunities to learn about the life that exists upon it. At certain places, whether they be windblown deserts, the ocean's depths, advanced laboratories, or mountaintop observatories, Earth also offers clues as to the prospects for life beyond it. Wherever the location, I found the story remained the same: Teams of scientists were at work, each passionate about their research, each asking what we can learn about life beyond Earth without ever leaving our home planet.

I decided there was a story to be told, celebrating their endeavors and joining them in the great adventure of performing science in an epic setting. However, for you the reader, I also wanted this book to act as an encouragement, a launch point for your own astrobiological adventures. So I have purposefully tried to visit locations that remain within reach of an average person with a passion for science—although I do admit to including a few places that offer more of a challenge. In this book I wanted to experience as many of these locations as possible, both as a professional scientist seeking to learn what modern astrobiology looks like in the field and as a guide, leaving you the reader a trail of crumbs to follow, with all the peril and pleasure of discovering things for yourself.

Some might call this scientific tourism—and I am happy for them to do so, as it is a pursuit I engage in myself and enjoy encouraging in others. Yet whether we engage in scientific tourism or professional research, the goal is the same: the acquisition of knowledge, be it personal or collective.

So here's to your own adventures, and I hope you enjoy reading about mine.

1
THE PALE BLUE DATA POINT

On February 14, 1990, planet Earth received a very special Valentine's Day greeting. The *Voyager 1* spacecraft, heading on a trajectory that would take it beyond the solar system, turned around and took one last look at home. It captured an image of Earth, barely a few pixels across, that showed our planet apparently suspended in a shaft of sunlight and surrounded by the black night of space. For one instant, one poignant orbital moment, all of Earth, all of humanity, was preserved in an image of deep emotive power. The idea for this parting shot was first proposed by Carl Sagan, and on seeing the image for the first time he was moved to reflect upon the fragile and precious nature of Earth as captured by *Voyager*—his "pale blue dot."

> Look again at that dot. That's here. That's home. That's us. On it everyone you love, everyone you know, everyone you ever heard of, every human being who ever was, lived out their lives. The aggregate of our joy and suffering, thousands of

confident religions, ideologies, and economic doctrines, every hunter and forager, every hero and coward, every creator and destroyer of civilization, every king and peasant, every young couple in love, every mother and father, hopeful child, inventor and explorer, every teacher of morals, every corrupt politician, every "superstar," every "supreme leader," every saint and sinner in the history of our species lived there—on a mote of dust suspended in a sunbeam.

Each generation on Earth faces its own particular challenges, and Sagan's meditations upon this image, with its implications for humanity, are just as relevant today—perhaps even more so—as when he first expressed them over thirty years ago.

One reason why the image of the pale blue dot is so striking is because it appears to show life on Earth existing in stark contrast to the dark void of space. But is our Earth really a lonely outpost of life in a bleak cosmos, or might it instead represent just one haven among many—a collection of pale blue pixels that form a single dot in a universal pattern of life? This book tells the story of how we are trying to answer that single, utterly profound question. It is the story of astrobiology—the scientific search for life in the universe—and, as we shall learn as we explore together, that search is dominated by the need to solve two very practical problems: Where should we look, and what should we look for?

This is a timely moment to consider these questions because the science of astrobiology is rapidly approaching its moment of first contact with alien life. It might come from strands of alien DNA sampled from an ocean locked within the ice moons of Jupiter or Saturn or ancient fossils preserved in the mineral matrix of a Martian rock. We will also consider whether a glimmer of light passing through the atmosphere of a distant exoplanet might be displaced by a molecule of oxygen photosynthesized under an alien sun or if we will discover, untouched in the vault of a distant asteroid, ancient amino acids that form the origin story of life on Earth.

Most astrobiologists—the scientists who are out there doing the looking—would freely admit to taking the optimistic view that alien life is indeed "out there." Yet those very same astrobiologists would also tell you that the privilege of being among the first generations of scientists to perform meaningful searches for life brings with it a compelling responsibility to search with precision and rigor. This book will tell the story of exactly where we are looking for life and why. We will learn why the planet Mars has long called to those searching for life; we will discover which among the small icy moons of the outer solar system offer the most beguiling hints that they may harbor oceans of life; and we will come to realize why, among the seemingly numberless stars of the night sky, astrobiologists are counting on the prospects of small, rocky worlds in orbit about dim red stars as the best current bet for discovering life beyond our solar system.

If it seems to you that much of the *astro* in astrobiology is concerned with deciding where to look, then you will not be surprised to learn that ideas drawn from our own *biology* are fundamental to deciding what to look for. We could focus on what life is made of (at least on our own pale blue dot): atoms of carbon, nitrogen, oxygen, and so on, ordered into remarkable chemical structures within the microscopic haven of a cell. Then again, we could consider what life does—at least insofar as nonliving matter does not do—by looking at how cellular metabolism differs from mere chemistry or how cell growth and division buck the universal laws of decay and disorder. Having cast planet Earth as a pale blue dot suspended in a shaft of sunlight, this book will consider a related idea—and only a slightly irreverent one: that of the pale blue data point. We will explore what Earth itself can teach us about the search for alien life, and our quest for answers will take us on five journeys across our home planet. Some of these journeys will show us how astrobiologists are using Earth as a training ground for missions beyond it. Others will take us to the spacecraft control rooms and telescopes where actual searches for life are being directed. In undertaking all these journeys I wanted to provide a personal answer to the question of what an astrobiologist actually does and to demonstrate, in the clearest possible manner, what the present-day science of astrobiology looks like in the field. I wanted to visit locations that not only reflect modern astrobiology in action but also offer you the reader the chance to get your own boots on the ground and retrace some of the same journeys that I was fortunate enough to make. We will ultimately learn that astrobiologists study life in the universe and that, whether we like it or not, planet Earth provides us with a solitary data point, pale blue in color, that represents—at least for today—our single example of life, wondrous in all its diverse forms.

One of the key ideas conveyed by the pale blue data point is that planet Earth provides a training ground for astrobiologists—an analogue of environments we might encounter in our exploration of the solar system. These analogues are the closest approximations to off-world environments that we can study without leaving home. And for those mission planners preparing to send the next generation of space probes to Mars, Jupiter, and beyond, these terrestrial analogues offer aspiring astrobiologists important lessons that we would do well to heed as we dispatch our robotic explorers in search of life.

Some of the most compelling environments in which to begin our search for life beyond Earth are found beneath the icy crusts of solar system moons such as Europa and Enceladus. These deeply mysterious moons orbit the giant gas planets Jupiter and Saturn respectively, and as we will discover on our journey they are thought to harbor vast oceans beneath their icy surfaces. Yet what can we, with our Earth-bound perspective, hope to learn about their hidden depths? The search for terrestrial analogues to these Jovian moons will, perhaps not surprisingly, take us to the depths of Earth's own oceans—to deep-sea hydrothermal vents some two kilometers beneath the surface of the northeast Pacific Ocean. These deep ocean hot springs host some of the most diverse yet utterly strange habitats for life found anywhere on Earth—whether above or below the surface. As a guide to what to expect in the profoundly alien depths of a Europan or Enceladan ocean, we could not ask for a better start to our travels.

The planet Mars has long held the hopes of astrobiologists in a beguiling embrace. We will learn about the first, flawed observations of illusory canals on Mars, an early example of technosignatures of life that perhaps told us more about human fallibility than about alien life. Yet Mars today is the most explored location beyond Earth—more robotic space probes have orbited, landed on, and rolled over Mars than any other planet or moon in our solar system. The search for life, or at least for the conditions that might support life, has formed a central part of almost all such missions. Where on Earth might we discover a terrestrial analogue of the red planet? The answer turns out to relate more to "when" than to "where," as we will learn that, some 3.5 billion years ago, Mars and Earth were warm, wet planetary twins. Given that life emerged on Earth in this distant eon, it is natural to ask if it could also have arisen on Mars. To answer this question we need to learn where on Mars might offer the best chance to discover—and recognize—any preserved traces of ancient life. Our own journey will therefore take us to remote rock outcrops in Western Australia where fossil evidence of the earliest life on Earth provides a direct example of what we might soon discover on Mars with a new generation of robotic rovers (ROVs) and sample-return missions.

The idea of the pale blue data point includes more than just the search for Earth-based analogues that help us prepare for the search for life in our solar system. In the spirit of exploring the astrobiology we can undertake without leaving our home planet, I wanted to include some stories where the universe comes to us—whether in the form of a twinkling stream of photons from a distant extrasolar planet or a meteorite recently fallen to Earth.

The past thirty years have witnessed a golden generation of exoplanet discovery. Not only have astronomers revealed an astonishing variety of exoworlds, but their abundance is such that we can now say with confidence that essentially every star in the night sky is likely accompanied by at least one planetary companion. This age of discovery has been driven by decades of innovation in telescopes and their cameras. What is it like to spend a night at a state-of-the-art observatory, to bear scientific witness to the arc of the firmament? What new telescopes, some currently under construction, will take us on the next step of the journey? For that step is a big one. We have so far discovered some thousands of exoplanets and rated them in terms of their *habitability*—the extent to which they offer potentially clement conditions that might support life. However, a new generation of giant telescopes offers the possibility to observe these planets and determine whether they are *inhabited*—to ask and answer directly whether they actually host life. In answering these questions I will retrace the steps of my life to my earlier career as an astronomer in Chile, where mountaintop observatories were my home and the stars of the night sky my constant companions.

The more we have learned about exoplanets, the more we have come to contemplate the history of our own solar system. Where did our life-bearing planet come from? If we rewound a movie of the solar system's history to its beginnings, would we have picked out the proto-Earth as fated for a special future? That's an opportune turn of phrase because some of the stills from the opening scenes of that movie can still be viewed today. They exist as meteorites, and each fused rocky fragment provides us with a snapshot from the earliest home movies of our own family of planets. Meteorites and the asteroids from whence they came allow us to determine with uncommon certainty the source of the chemical raw materials that were later brewed up in Earth's cauldron of life. We have studied the earliest fossils of life on Earth, but meteorites offer us the chance to venture much further back in the story of our planet: They are the relics of planetary formation, fragments of the solar system that have floated untouched for over 4.5 billion years. By falling to Earth, meteorites present scientists with free samples of solar system material delivered (almost) to their door. To find them we will travel to the barren stony plains on the periphery of the Sahara desert and walk the ancient paths of nomadic herders in search of cosmic clues that we can hold before our own eyes. However, holding meteorites in the palms of our hands will not mark the end of this story. We will trace their blazing trails back to the asteroids from which they fell, traveling along with the Japanese and US space missions daring to sample fragments from their dark surfaces and return their precious grains of knowledge back to Earth.

The last of my journeys could be considered the most whimsical—it was certainly among the most interactive—as I was fortunate enough to join a team of dolphin researchers in the waters off the Bahamas for what could only be said to be a tangential effort at interspecies, let alone alien, communication. Yet the idea of using giant radio telescopes to eavesdrop on alien conversations, or even participate in them, was the intellectual big bang moment that initiated the modern science of astrobiology. As we shall see, it was not for nothing that the early astrobiology pioneers named themselves the Order of the Dolphin.

Why did I choose these locations to highlight modern astrobiology in action rather than any of a host of others? Each of these journeys will bring us to the cusp of new discoveries in the precious moments before they are made. I cannot tell which, if any, of these endeavors—the search for life on Mars or Enceladus, the twinkle of light from an exoplanet, or the faint whiff of chemical insight gained from a few grams of asteroidal material—will provide the next clue or even the smoking gun of a clear detection of life. But among all the stories I could have chosen, these represent some of the best bets that we are currently working on.

If I neglect to touch upon your favorite astrobiological project, do not think that this reflects any lack of enthusiasm on my part. Might the detection of phosphine in the atmosphere of Venus point to the strange, but not unguessed at, existence of life among its clouds of sulfuric acid? In a similar vein, the prospect for NASA's *Dragonfly* mission to detect the building blocks of an alien ecosystem on Saturn's moon Titan offers a stark challenge to our Earth-based ideas of life as we know it. Great ideas abound. However, it is not my aim to provide you with an exhaustive account of the breadth of modern astrobiology. There would simply be too many stories to include and the resultant book (and surely its readers) would groan under the weight of the included material. I hope instead to achieve a balance between giving you not just a description of astrobiology as a science, but also a very real sense of what astrobiology looks like in the field.

In my earlier book, *All These Worlds Are Yours*, I told a number of astrobiological stories. Yet this time around I wanted to take you out of the classroom and into the field to celebrate some of the special places on Earth where one can get up close and personal with locations rich in scientific mystery and revelation. We exist on a fascinating planet, and we should never miss a chance to be reminded of this fact. You might find yourself at a lonely outcrop of rock where you can reach out and touch the subtle imprint of life from Earth's distant past or within the reverent interior of a telescope dome as it bears passive witness to remote, starlit worlds. Each experience is akin to entering a scientific cathedral, evoking as it does a special sense of grandeur as you are confronted by the immensity of time and space. If this new book can convey to you even a small part of that experience, then my task will be largely complete.

Why should we search for alien life? What could we possibly hope to gain? It is a common theme in science, as it perhaps is for life in general, that the simplest questions turn out to be the most difficult to answer. From my own perspective, the motivation that drives each scientist to search for alien life is not so very different from the curiosity that draws a member of the public to attend a talk describing the latest astrobiological discoveries (or to buy a book about them). The difference is one of degree, yet it would be an unwise scientist who claimed any special authority in speculating on why we should search for alien life. Instead, I offer here two personal perspectives for your consideration—one view takes in the grand philosophical prospect of learning of our place in the wider, living cosmos; the other considers the small-scale wonder that might come from observing an alien microcosm in a drop of Europan water.

On the grandest of scales, one can argue that human discovery has led us on an outward-looking journey throughout our history. That statement sounds very fancy, yet when we look closer at the impact those scientific discoveries have had on us as human beings, that grand journey turns out to be a lot more humbling than you might first have thought. In 1543 the mathematician and Catholic canon Nicolaus Copernicus published what would prove to be a revolutionary book—for once the pun is not intended. *On the Revolutions of the Celestial Spheres* presented a mathematical model in which the sun, instead of the Earth, was at the center of the solar system. Earth was cast adrift as just one planet among the other known worlds. But if the

Earth could be apparently demoted in this way, what might this mean for our sun? A subtle yet profound idea had been inadvertently revealed. Could the stars that studded the celestial sphere be suns in their own right, dimmed to the faintest of twinkles only by virtue of their distance? It was a compelling thought that opened up a universe of possibility. If we followed it to its logical conclusion, would not those distant suns be accompanied by planets of their own?

In 1584 Giordano Bruno, a philosopher and Dominican friar, followed exactly this line of thinking when he famously speculated that we were just one among many possible worlds in orbit about the stars. A short excerpt from his work *On the Infinite Universe and Worlds* reveals a remarkably modern mind at work. During their extended dialogue, the inquiring Elphinor asks the philosopher Philotheo, "Why then do we not see the other bright bodies which are earths circling around the bright bodies which are suns?" To which Philotheo offers the strikingly prescient reply (which I paraphrase), "We do not discern the earths from the larger suns because, being much smaller, they are invisible to us." Bruno had opened the gates of human imagination to the idea that became known as the plurality of worlds—that our universe was filled with a plenitude of planetary possibilities. His clear statement of the problem of discerning a faint planet from the glare of its parent star would remain the single yet all-encompassing challenge faced by exoplanet-hunting astronomers for the next four centuries. The story of how we overcame it contains its fair share of both triumph and tragedy, and it is one that we will return to in earnest later in our journey together.

More cosmic revelations were to come, and with the benefit of our imaginations we can zoom out from the solar system to reveal our place in the outlying stellar neighborhoods of the Milky Way galaxy. Through the 1920s, Edwin Hubble and other astronomers discovered that the universe beyond the Milky Way was populated not by individual stars but instead by great cities of stars. Our own Milky Way was revealed to be just one galaxy among a staggering multitude. It turned out not that we were simply in orbit about one sun among hundreds of billions of stars in the Milky Way, but that our galactic home was one of hundreds of billions of galaxies within the observable universe.

Hundreds of billions of galaxies, each populated with hundreds of billions of stars might well mean that we exist in a universe populated by a number of planets best written down as the number 1 followed by 22 zeros. Though we can calculate that number, it stubbornly defies our human comprehension. The universe had become larger than we could have possibly imagined, and without a good dose of cosmic humility we risked losing sight of our place within it.

Astronomers are not the only visionaries who have challenged our place in the cosmos. Closer to home, Charles Darwin was among the first to realize that humans do not occupy a privileged position compared to the rest of life on Earth. His theory of evolution through natural selection offered humanity a daring scientific vision—one in which all present-day life on Earth had evolved from simpler forms. When viewed across the vast loom of time, life on Earth forms a woven fabric of existence whose warp and weft have grown through at least 3.5 billion years of our planet's history. It remains an astonishing conclusion, one that reveals the depths of time that have existed prior to humanity's brief appearance on the stage of life and that is rivaled only by the accompanying vision that we humans represent merely the tiniest of specks within the vastness of the universe.

Whose name will we associate with the discovery that life on Earth is only one example among many—a single tree of life within a larger forest of possibilities? What will we learn when we travel beyond Earth and interact with these life forms? Darwin, Hubble, and Copernicus challenged humanity with their scientific vision and redefined our relationship with nature and the universe. The discovery of alien life will certainly offer a further challenge to how we view our place within the cosmos. Though we don't know who will make this discovery or when, the scientific stories we will encounter in this book will allow us to speculate on where and how that discovery might occur. How will we react? Will we gasp with awe and wonder or recoil in skepticism and denial? Perhaps it is a fool's errand to anticipate the Babel of reactions that will greet any claimed discovery of alien life and try to impose some kind of scholarly authority upon them. Yet if the claim is based upon scientific evidence, and if it supports the weight of scrutiny applied to it, that moment of discovery will change our view of the cosmos and our place within it—whether we want it to or not. Our perspective will have shifted to take in a dizzying cosmic view, and the world as we perceive it will never be the same again.

So much for the grand visions of our place in the cosmos. When searching for alien life there is another, perhaps simpler, pleasure to be taken, one that considers the microcosms that might lie within a drop of water. In the late 1600s, onetime Dutch cloth merchant and lensmaker Antonie van Leeuwenhoek was the first person to glimpse these strange and alien worlds using his new invention, the microscope. Through it he saw tiny worlds, complete in every detail, of single-celled creatures suspended in a drop of water—a perfect miniature canvas of life to stand in comparison to *Voyager*'s later image of a microscopic Earth suspended in a shaft of sunlight.

The technology he used was simple and effective—a glass bead mounted on a handheld brass fixture—so simple in fact that it could have been invented at any point during the preceding three thousand years of human glassworking. It is a wonderful example of a scientific "what if" moment—a chance to ponder how human thought might have reacted to an earlier discovery of the existence of microscopic living worlds not just under our noses but up them as well. As it was, following van Leeuwenhoek's observations of the microcosmos, our own perception of the world grew larger. Even here, though, history offers us an enlightening example of how claims of new life are met: The Royal Society in England, to whom van Leeuwenhoek communicated his results, felt compelled to dispatch a delegation of trusted scientists to the Netherlands to verify his extraordinary claims. From the point of view of the scientific establishment of the time, believing definitely required seeing.

As science later journeyed from microscopy to microbiology, we realized not just that we humans were also composed of such tiny cells but that all life on Earth shares a fundamental chemical kinship: the central role of DNA as an evolutionary molecule, the use of the molecule adenosine triphosphate (better known as ATP) as the currency of energy transactions, and, at the level of individual atoms, the building blocks of carbon, hydrogen, nitrogen, oxygen, phosphorus, and sulfur. The totality of life on Earth, astonishingly diverse as it is, is built upon a deep and shared chemical foundation.

Yet viewing a microcosmos in a drop of water invites us to ask what other worlds we might discover in a drop of alien water and what pattern of life such strange organisms might follow. The icy moons of the outer solar system, chief among them Europa and Enceladus, currently present the most compelling evidence for liquid water beyond the confines of Earth. The *Cassini* space probe flew through the geyser-like plumes of Enceladus in 2008, and its onboard instruments detected the presence of salt, water, and a slew of undeciphered carbon-based compounds. Part of my personal motivation to discover alien life therefore comes from the desire to place a drop of Europan or Enceladan water before my microscope and to ask simple yet fundamental questions of any life it contains. Are there cells present, and how are they constructed? What atoms do they use, and in what arrangements? What is the recipe followed by this new example of life, and what can we learn about our own recipe in comparison with it? Back home on our pale blue data point, our journeys to visit deep ocean hydrothermal vents will offer us a compelling environment in which to hone our scientific skills as we plan our return to these alien oceans—both cryptic and enthralling in equal measure.

At this point we might be wise to consider applying the scientific equivalent of a sanity check: If we are so confident that alien life is out there, just waiting to be discovered, perhaps a whisper of caution might warn us to first make sure that we are able detect the life we know exists here on Earth. It was with this cautionary idea in mind that in 1990—the same year that we were using *Voyager 1* to take our long-distance selfie—the *Galileo* space probe was tasked with giving us a much closer view of our home planet in order to answer what appeared to be a rather quirky rhetorical question. Could we use a space probe originally designed to study the planet Jupiter to detect life on Earth? On October 18, 1989, the space shuttle *Atlantis* launched the *Galileo* mission on the first stage of its circuitous six-year trajectory to Jupiter. To save propellant en route, *Galileo* executed three gravitational assist maneuvers, performing a grazing flyby of Venus and two of Earth to pick up additional velocity for its journey to the outer solar system. Carl Sagan recognized that these Earth flybys offered a unique opportunity to use *Galileo*'s instruments to survey our home planet with a view to detecting the signatures of life. In principle, it sounds easy. After all, in 1990 there were 5.3 billion people living on planet Earth—How could you miss them? But what criteria would you define in advance to determine whether life is present? Many of us fall back on "I'll know it when I see it," but that serves us poorly as scientists. Sagan's argument could therefore be paraphrased as, "If you recognize life when you see it, then what do see when you look at Earth?"

The idea of using the entirety of planet Earth to perform an astrobiological experiment is perhaps the most ambitious application of the idea of the pale blue data point that we will come across in this book. With the *Galileo* flyby we in effect put ourselves in the place of an alien explorer who has traveled to Earth in search of life. Having crossed the gulfs of space to arrive at this view of a pale blue dot in orbit around an unremarkable star, what telltale clues would point to Earth as being a living planet? What *Galileo* ultimately observed on its passing flyby of Earth was the spectroscopic signature of abundant oxygen and methane in our atmosphere and the surface absorption of sunlight due to molecules of chlorophyll. In this respect *Galileo* had discovered Earth life at its most fundamental. It had detected the planetary metabolic output of some of the most basic creatures on Earth—microbes and plants—in addition to the spectral trace of one of our planet's most important biological molecules. The fleeting trace of humanity glimpsed by *Galileo* on its flyby was that of our radio broadcasts. Our planet appeared to "chirp" at radio wavelengths in a way that was very different to the bursts of natural emission associated with atmospheric lightning. Each TV and radio station created its own signature of electromagnetic ripples, carrying the sounds of humanity out into space like so many stones carelessly tossed into the deep waters of the cosmos.

These observations—of abundant atmospheric gases such as oxygen and methane, surface absorption from chlorophyll-like molecules, and radio emission from broadcasting inhabitants—have been labeled the "Sagan criteria" for life detection. This is an unfair term, as it perhaps gives the impression that these are exhaustive criteria, encompassing the totality of signatures expected from life (something that Sagan never intended they should be). The true importance of the observations conducted by the *Galileo* probe was to highlight the idea of biomarkers—detectable phenomena that point unambiguously to the action of life.

Two of the most enduring of the *Galileo* biomarkers were the observations of oxygen and methane in our atmosphere. It wasn't just that oxygen and methane were present, as these two molecules are found to a greater or lesser extent in the atmospheres of all the planets of our solar system. It was that the abundance of each molecule in our atmosphere was in marked excess compared to what you would expect to find on a nonliving Earth.

The key idea here is that of planetary chemical equilibrium, and it was first put forth by the pioneering chemist James Lovelock in 1972. He called it the Gaia hypothesis, and it explains that Earth's biological inhabitants are linked by a multitude of chemical reactions to our planet's atmosphere, oceans, exterior crust, and molten interior. In this great connected system every terrestrial molecule is in chemical balance with every other in a finely tuned planetary equilibrium. If the balance is tipped, then the chemical equilibrium of the planet will change in response. Take the example of atmospheric methane. Without the action of biology, in this case methane-producing microbes that break down organic matter, almost all of the methane in our atmosphere would be broken apart by sunlight within approximately twelve years. No other effect on our planet would replace it at the level at which it currently exists.

The *Galileo* spacecraft's observations of planet Earth during its 1990 encounter helped fix in astronomers' minds the idea of searching for exoplanet atmospheres that showed clear biomarkers—atmospheres where the scales of the chemical balance were tipped firmly in favor of life. Our journey to encounter these worlds will take us to some of the largest telescopes both on Earth and above it as they attempt to capture the first faint traces of exoplanet atmospheres. The pale blue data point in this case will be our vantage point from which we look out to distant worlds, and, among the scientific cathedrals we will encounter on our travels, these giant observatories will be some of the largest to be built by human hands.

Ultimately, the *Galileo* experiment provided a snapshot of what extraterrestrial astrobiological data might one day look like—mimicking what we hope to find when we glimpse a pale exoplanetary dot through the aperture of a giant telescope. Perhaps the most important legacy of the *Galileo* flyby is to remind us that any single measurement rarely provides the whole picture that one might have hoped to obtain. In this sense scientific discovery in general—and astrobiology in particular—is akin to attempting to solve a particularly challenging jigsaw puzzle. Except in this case you are given just a few pieces, you are not told how many pieces there are in total, and the picture on the front of the box may bear little resemblance to the image you finally put together!

The *Galileo* mission taught us how we might recognize life from the perspective of a planetary flyby. Now, as we return our gaze to Mars, Europa, and beyond, we need to consider how we might land a robotic spacecraft in these locations and how we might recognize biomarkers at the molecular level via experiments performed in an onboard laboratory. For even as astrobiology approaches its first detection of life, scientists are still grappling with a fundamental challenge: How do you define a scientific measurement that will confirm—or refute—the existence of life in a physical sample?

Cogito, ergo sum—I think, therefore I am. So reasoned René Descartes in response to the troubling question, Am I a living, conscious being? Putting aside for one moment the nature of our own consciousness, can we attempt to answer a related question: What measurable phenomena can be used to define life? One philosophical challenge we face as astrobiologists is that there exists no measurement unit for life. Life is not a quantity, it is a process—a chain of linked chemical reactions occurring within the physical space enclosed by a tiny cell. The distinction is important because the life-detection experiments we will encounter on our travels will be based upon remotely collected measurements from soil samples, rock composition, and the chemical makeup of particular gases. The challenge of recognizing the presence of living organisms using such measurements is that they represent a zoomed-in, close-up view of the basic processes associated with life, and we will not be able to fall back upon the broad, holistic view of life available when we view its totality. This is perhaps one of the more obvious places where the idea of the pale blue data point comes into play. Life on Earth provides what is currently our only example of life in the universe. To ask what is life at its most simple, we must answer the question, What does life do that nonliving matter does not? If we further take the point of view that biology is simply chemistry made animate, then we must answer this question at the biochemical level.

On our journey together we will encounter a number of scientific definitions of life. However varied their application, each falls into one of a small group of definitions: I metabolize, therefore I am; I am ordered, therefore I am; I evolve, therefore I am. (Should you detect an extraterrestrial signal from intelligent, communicating aliens, you might also use, "I am impressed, therefore they are.") Each of these definitions concerns one particular facet of cellular life: the cyclic processing of chemical energy, the physical structure of the cell, and the process of molecular reproduction with small random errors. You might well ask which of these definitions is the most important. Must the blob of goo at the end of our microscope satisfy all three definitions—or more—to be considered a living organism? Is our definition of life too broad in scope or too narrow—or is it perhaps just right? To answer that question we may have to introduce some flexibility into our thinking.

When defining whether an entity is living or not, we like to think that it would be either 100 percent living and 0 percent nonliving or vice versa. However, by thinking about life in this way, we allow ourselves no middle ground through which chemistry and order might stumble forward together in a continuous journey toward life. What if instead we allowed ourselves the leeway to wander between these extremes of black and white, using shades of gray to capture the nuances and complexities of life as a chemical continuum? The simplest step we could take would be to give a sample under analysis a grade between 0 and 100 in a manner akin to a high school biology class. Zero would represent the consummate failing grade of inanimate matter. A score of 100 would represent an organism that shares all of the common properties of modern Earth life—order, metabolism, and the capacity to evolve.

Any organism displaying just one of these traits could be given a grade of 33/100. But, though eminently practical, this approach raises other questions that remain stubbornly harder to answer. What is the passing grade needed to be considered a living organism? If modern life represents a perfect score, where would protolife sit on this scale? I have assumed that we would score order, metabolism, and evolution equally, but we must recognize that they might not have appeared at the same time in the story of life on Earth—perhaps cellular order appeared first and provided a safe haven in which cyclic metabolic reactions could develop. Other consequences of this way of thinking lead us to contemplate the potentially bizarre, for if modern life represents a score of 100/100, then might we one day encounter an organism that scores more than 100 on our test of life? That would be an interesting encounter—and likely a rather humbling one as well.

Despite its pitfalls, assigning a grade to life allows us to approach such biological quandaries as whether a virus is alive or not. For many, a virus remains a nonliving entity, condemned for lacking the molecular mechanism to copy its own DNA, despite exhibiting metabolic activity and cellular order. Assigning equal weight to metabolism, order, and (flawed) reproduction, we would give a virus a grade of 66/100. The important point is that, when trying to answer the question of what life is, the idea of a flexible boundary between living and nonliving entities provides us with just enough wiggle room to consider chemical systems (it seems biased to call them organisms) that display some, though not all, of the elements of life as we define it today.

Putting the philosophy of this approach to one side, the discovery of an alien entity satisfying just one or two of these tests would represent a moment of deep scientific revelation. However, in practice science would likely demand that a series of tests based upon each of these three biochemical phenomena pass their respective thresholds to be considered a definitive detection of life. Fortunately the history of astrobiology is replete with examples where these definitions have been applied with either cool caution or willful abandon. Each serves as the astrobiological equivalent of an ancient fable, warning of the perils that adhere to claims of alien life, so that we might learn wisdom as we embark upon searches of our own.

At the turn of the twentieth century, the astronomer Percival Lowell, upon viewing what he perceived to be vast, engineered canals on the surface of Mars, effectively announced, "They dig, therefore they are!" Unfortunately for Lowell, the scientists who refuted his claims using telescopes of increased collecting power and angular resolution were led to retort, "He is misled, therefore they are not." (Carl Sagan was moved to reflect that even this episode provided a definition of life; it just happened to be at the wrong end of the telescope.)

More sober was the definition of life applied by the scientists who constructed the biological experiments carried by *Viking 1* and *Viking 2*, the first space probes to land on Mars during the bicentennial summer of 1976. These landers would test for the presence of living organisms by delivering scoopfuls of Martian regolith to an onboard laboratory, and three out of the four experiments undertaken by the *Viking* landers aimed to detect the metabolic action of Martian microbes. One test in particular, the labeled release experiment, fed a small dose of liquid nutrients to a sample of Martian soil. When a steady plume of gases was released from the soil—a result that appeared consistent with microscopic Martian life metabolizing the supplied nutrients—the NASA team running the experiment had to step back and consider carefully whether they had really detected Martian life or not.

On closer inspection the result turned out not to be as clear-cut as it first appeared, and *Viking* provided two important lessons in how one should interpret life-detection experiments: First, one requires multiple positives to believe a detection of life. Three of the four *Viking* experiments were designed to test for different aspects of metabolic activity. The fact that one of them could be called successful might lead us to offer an initial grade of 33/100 in our search for life. However, the fourth test, the gas chromatograph mass spectrometer measurements, failed to detect in the samples of Martian soil any volatile carbon-bearing molecules that might have been associated with life.

That quandary led the *Viking* science team to look again at the labeled release experiment and to consider whether any nonliving chemical processes could mimic their result. This is the second important point to consider when interpreting the results of a single experimental result—if your knowledge of the Martian soil chemistry is incomplete, then your conclusions will also be incomplete. In this case the presence of perchlorate salts detected in the Martian soil by the 2008 *Phoenix* lander apparently explained how the *Viking* results could have been produced by a nonbiological oxidation reaction. Viewed in hindsight, the *Viking* biology experiments conducted on the surface of Mars provided valuable but unexpected lessons in how to search for life. However, only eight years after that glorious summer of 1976, a completely unexpected opportunity to learn about life on Mars quite literally dropped into the lap of a new generation of NASA scientists.

In the austral summer of 1984, Roberta Score, a member of the US Antarctic Search for Meteorites team, spotted a conspicuous black rock from her snowmobile as she traversed the otherwise blank expanse of the Alan Hills region of western Antarctica. Small enough to fit in cupped hands, the meteorite—identified by its fractured black fusion crust—was bagged, tagged, and kept in cold storage for its long journey to its new home at the Antarctic Meteorite Laboratory of NASA's Lyndon B. Johnson Space Center in Houston, Texas. It would take the best part of a decade for scientists to fully appreciate the rare gift that the solar system had fortuitously delivered to Earth. The meteorite tagged as ALH84001 would turn out to be a chunk of Mars blasted off the planet's surface by a meteor impact some seventeen million years ago. It had spent millions of years wandering through the inner solar system before making its fiery descent to Antarctica thirteen thousand years ago. However, the rock itself was far older than these dates, with a precise analysis of rare radioactive isotopes in the sample indicating an age of 4 billion years. That meant that ALH84001 was formed during the era of Martian geological history known as the Noachian, or "wet Mars," period. When we travel to the ancient rocky outcrops of Western Australia, we will see for ourselves that life was present in shallow marine environments on Earth some 3.5 billion years ago. ALH84001 therefore offered scientists at the Johnson Space Center the opportunity to look for signs of life in similar conditions on early Mars—if only they knew what to look for.

David McKay made his start in scientific life as a geologist recruited by NASA to support the *Apollo* program. He trained both Neil Armstrong and Buzz Aldrin to prospect on the moon, and he was the only geologist present at mission control for the astronaut's walk on the lunar surface. By the early 1990s McKay was one of NASA's most experienced investigators of extraterrestrial material, whether in the form of lunar samples, cosmic dust particles, or—fatefully as it would turn out—meteorites. McKay and his team of collaborators discovered in ALH84001 a treasure trove of Martian history. The rock had formed some 4 billion years ago and was later exposed to liquid water that, upon evaporating, had deposited a patina of carbonate molecules in its fractured interior. Complex carbon molecules, in the form of polycyclic aromatic hydrocarbons, were also found on the interior surfaces of the meteorite. It was starting to look to the research team that ALH84001 contained all of the raw materials for life. But what about traces of life itself? No one expected to discover a living organism in the sample,

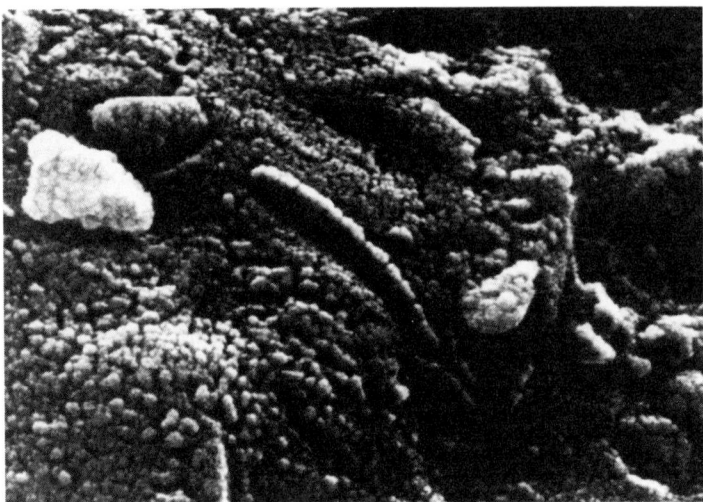

ALH84001 A fossil Martian microbe or merely a quirky mineral grain? This electron microscope image of the interior of the Martian meteorite ALH84001 purports to show a bacterium-like organism some two hundred nanometers long—approximately five hundred times less than the thickness of a human hair. Credit: NASA, https://web.archive.org/web/20051218192636/http://curator.jsc.nasa.gov/antmet/marsmets/alh84001/ALH84001-EM1.htm, Public Domain, https://commons.wikimedia.org/w/index.php?curid=229231, https://en.wikipedia.org/wiki/Nanobacterium#/media/File:ALH84001_structures.jpg.

but what about preserved traces of life akin to the ancient fossils that marked the earliest evidence of life on Earth?

The critical clue discovered by McKay and his team were microscopic crystals of the mineral magnetite. On Earth magnetite is a familiar find at mineral and gem shows, yet it is also produced on microscopic scales by microbes in order to turn themselves into cell-sized compass needles that align with Earth's magnetic field. The tiny magnetite crystals in ALH84001 displayed a particular crystal structure that, at least initially, displayed a close match to those produced by microbes on Earth. However, the more photogenic clue, the one that would eventually appear on the front cover of the prestigious academic journal *Science*, was of tiny, sluglike mineral deposits that bore an uncanny resemblance to terrestrial bacteria.

McKay and his team assembled these clues to make a case for ancient, fossilized life in ALH84001. Once their article had been accepted by *Science*, the news was placed under a strict embargo. However, as the journal's publication date approached, news began to leak out, and NASA organized a hurried press conference in an attempt to regain the initiative. Thus began a frenzied cycle of headlines that spun the researchers all the way to the White House lawn to stand behind President Bill Clinton as he informed the nation of the potential impact of this discovery.

But exactly what had been found, and what grade would we give to the claims of McKay and his team? Taken at face value, the microbe-like structures and magnetite crystals both provided evidence of cellular order. This might lead us to assign ALH84001 a starting grade of 33/100. Although there was no evidence of metabolic activity or of evolutionary molecules, you might be hard-pressed to find any after almost 4 billion years. For that reason, McKay and his team essentially asserted that the features they had discovered were so unlikely to have been produced by a nonliving source that they provided sufficient proof of life on their own.

Once the news of ALH84001 had broken, the wider scientific community was able to provide its own assessment of the evidence—and the views were not favorable. Rather like a team of defense lawyers dissecting the prosecution's case point by point, the scientific community came up with alternative, nonbiological explanations to account for each piece of evidence extracted from the meteorite. Heated words were exchanged, and the tense atmosphere was not helped by the initial secrecy imposed on the journal article. McKay and his team, not unreasonably, felt pushed onto the defensive if not vilified for their claim.

Cooler heads, among them Baruch S. Blumberg, a Nobel Prize winner in physiology and medicine and the first director of NASA's Astrobiology Institute, realized that scientists were performing astrobiology in real time and that the community was learning by doing. He counseled David McKay and his team to focus less on the search for a smoking gun that showed unambiguous proof of life in ALH84001 and instead to take heart from the converging lines of evidence that suggested that more than mere chemistry was occurring in this tiny fragment of the red planet. Although this more sober take on ALH84001 hardly made headlines, this unsuspecting chunk of the planet Mars forced a generation of scientists to confront what they really considered to be evidence for alien life and what they did not.

The experience of the scientists who studied the *Viking* experiments and ALH84001 provided an important reality check to the nascent field of astrobiology. The hope of discovering clear evidence of life—that magical eureka moment—was replaced by the reality that scientists were more likely to uncover a network of clues, where the pattern of dots on the page, rather than any one hint by itself, would reveal the faint outline of life. In terms of our high school grading analogy, each time we explore a new environment in the search for life, we will attempt to build our score one experiment at a time, by combining lines of evidence drawn from chemistry, mineralogy, microbiology, and a host of other fields. Only then will we submit our results to the scientific community to see whether we have achieved a passing grade.

Our pale blue data point gives us a valuable opportunity to apply these ideas in practice, and the scientific environments we will encounter—ancient fossil landscapes and the living ecosystems of hydrothermal vents—will allow us to test our current ideas of what constitutes evidence of life. The burden of proof required to claim the discovery of life on Earth is clearly lower than that required to claim to have discovered life beyond it. However, if we can identify meaningful analogues of off-world environments here on planet Earth, then we can thoroughly test our ideas for detecting life before committing them to a particular space mission. In this manner, our home planet can provide us with some, though certainly not all, of the answers to the questions that concern us as astrobiologists. They are partial answers, to be sure, that carry with them an inherent flaw: Why should the alien organisms we seek resemble those of Earth? However, the principle in all new endeavors is that you have to start somewhere. And if astrobiology is a journey of a thousand miles, then learning to understand the significance of our home planet must surely be the first step.

Is anybody out there? This remains the ultimate question for astrobiologists, whether that "anyone" consists of microbes on Mars or aliens broadcasting on Tau Ceti. It is an intimidating question, profound in its import and, for much of our history, one that has appeared to be too large to answer—at least in one go. An alternative approach is to break down that one big question into bite-size chunks. This was the route proposed by Frank Drake when, as a young radio astronomer in 1961, he famously summarized the numerical factors that would play a role in calculating an answer to the related question, What is the *probability* of life existing beyond Earth? The Drake equation can be written as a series of symbols that represent the key physical terms involved: $N = R_* \times f_p \times n_e \times f_l \times f_i \times f_c \times L$. From left to right we see that the number of communicating civilizations in the Milky Way galaxy (N) is equal to the rate at which new stars are born (R_*) subsequently multiplied by the fraction of stars that host planets (f_p), the number of Earth-like (or more generally habitable) planets per parent star (n_e), the fraction of such worlds on which life develops (f_l), the fraction on which intelligent life evolves (f_i), the fraction that go on to communicate (f_c), and finally the average lifetime of such communicating civilizations (L). Although in its original form the Drake equation allows one to compute the *number* of worlds on which intelligent communicating aliens arise, to obtain a fraction or probability all one has to do is divide his answer by the total number of worlds available in the Milky Way.

In 1961 the only term in the Drake equation that was well understood was the rate at which new stars were born in the Milky Way. Every term to the right of this outpost of knowledge remained potentially important yet completely unknown. For some scientists, the inclusion of these unknown terms condemned the Drake equation to the realm of snigger-worthy speculation. In one sense this view is justified, as anyone who attempts to use the Drake equation to predict the actual number of communicating civilizations is essentially pulling numbers, like rabbits, out of a hat. But rather than using it as a turn-the-handle formula, Frank Drake always expressed the opinion that what was important about his eponymous equation was that it provided a clear statement of what we need to be thinking about next. In this sense the Drake equation resembles a scientific wish list. If you want to answer the big question of whether there is indeed anyone out there, the equation helps you to identify the only slightly less big ideas you have to learn about first.

When first written down in 1961, the second term in the equation—the fraction of stars with planets—was completely unknown and perhaps even seemed unknowable. As we shall discover on our journey to the astronomical observatories of northern Chile, it has taken scientists a generation, some thirty years give or take, since the discovery of the first planet in orbit about a relatively normal star, to determine this number. I don't think it gives away too much at this stage to note that that fraction is large—close enough to one to conclude that writing $f_p = 1$ in the Drake equation would not be too far from the truth.

Having made that statement, though, it seems appropriate to pause and consider its implications. It is astonishing that each and every star you care to look at in the night sky likely hosts at least one planet of its own, and the physical realities of the worlds we will discover on our journey will prove to be equally astonishing in their capacity to surprise us. However, within the broader context of the Drake equation, the fact that $f_p = 1$ is not only astonishing, it is also *encouraging*. It tells us that the fraction of stars hosting planets does not appear to be a limiting factor—a cosmic bottleneck if you will—and that our search for life can continue on its journey through the Drake equation.

What about the next number to the right, the next unknown? As written by Drake, n_E describes the number of Earth-like planets about each star. Our present knowledge of the science of planet formation is too sparse to calculate this number directly, largely because for so long we were limited to just one solar system, and only now are we starting to detect exoplanets in sufficient numbers to map out the physical principles involved. We will therefore take a different route to understanding this number and turn instead to the earliest history of our own solar system—as told to us by meteorites and a bold new generation of asteroid sample-return missions—in order to learn not just how our planet was assembled, but where the chemical raw materials that make up life on Earth were first synthesized.

The next term, the fraction of habitable worlds that actually go on to develop life, well, that's the big one. Determining that number is *the* goal of all present-day astrobiology, whether one is taking a relatively close-up look at the planets and moons of our solar system or the distant view that we currently have of exoplanets. The need to understand how life emerges on a habitable world will motivate our journeys to two of the strangest habitats on Earth. One is preserved in our planet's distant past and is revealed by fossils in Western Australia that are 3.5 billion years old. They will show us an unsettling, alien world where life first emerged along the shores of a lurid green ocean, under skies tinged a deep orange. Our other expedition will take us to the depths of Earth's oceans, where we will use robots to learn from the profoundly strange and equally alien ecosystems found at hydrothermal vents. Each of these journeys will take us as far as we can travel in our search for alien life in the solar system without actually leaving Earth. And what we learn at each location will provide us with critical insights as we contemplate the robotic exploration of the planet Mars and the icy moons of Jupiter and Saturn.

But what about the final terms in the Drake equation, those that consider the possibility that the Milky Way is populated with intelligent, communicating aliens? It is one thing to discuss the construction of a radio telescope and how it might be used to detect intelligent alien signals in our Milky Way galaxy, but it is quite another to contemplate deciphering any meaning that such a message might contain. Not surprisingly, our pale blue data point will offer us one last analogy to guide our speculations as we interact with dolphins, some of planet Earth's most compulsive communicators. Our journeys here will bring us to a meeting of minds and a profoundly humbling recognition of the fact that we are not the sole inhabitants of this pale blue data point to display both intelligence and the means to communicate it. There can perhaps be no more appropriate way to conclude our own very human survey of what life on Earth can teach us in the search for life beyond it.

In the more than sixty years since the Drake equation was first written down, no one has yet come up with a better way of summarizing the key questions to be answered by astrobiology. In this sense the durability of the Drake equation is a testament to its scientific utility. And you don't have to be looking for E.T. to make use of it. Should your ultimate goal be the discovery of even basic alien life, then the Drake equation (minus some of the later terms) would still be of use to you. My own opinion is that the Drake equation remains valuable today because it keeps us honest. It tells us how far we have come in our search for alien life and, dauntingly so, how far we may have yet to go. Perhaps most importantly of all, it tells us what questions we need to be asking next. This book is about asking those questions, and in doing so we will travel the length and breadth of our home planet in search of answers. The Drake equation will be our guidebook on our journey. Although it will not serve as a detailed road map to the discovery of alien life, at each of our stops en route to that destination, it will provide us with some much needed context as to how each small piece of the puzzle fits into the bigger picture.

Our story in this chapter began with a distant view of our home planet. Having taken that final image of Earth, huddled with the other planets in the fire-like glow of our sun, *Voyager 1* continued its journey into the outer solar system and beyond. For now, although we remain here on Earth, we can use our time as astrobiologists wisely. The scientific stories in this book all gaze outward, as we must do in order to discover life beyond Earth. We will look to Mars and the ocean-bearing moons of Jupiter and Saturn. We will marvel at the plenitude of exoplanets and treasure the fragments of our solar system that fall to Earth or are returned to us by space probes. In learning about the prospects for life to exist beyond Earth, we will also discover just how much planet Earth has to teach us about what evidence we might discover on our search and, just as importantly, how we might interpret those findings as scientists.

Approached in this way, the pale blue data point can help us answer a question of central relevance to astrobiology: If we treat the Earth as an alien planet, how would we confirm the existence and nature of the life it hosts? As long as we remain aware that Earth is just one planet, and if we can avoid the risk of casting our subsequent searches for alien life too firmly in Earth's own image, then exploring this pale blue dot will prepare us for our coming forays into the solar system and beyond. We can, of course, listen and watch from our vantage point of the pale blue Earth. It may well be that we detect the signature of alien life at a distance without ever leaving home. And indeed this will be our approach when we ascend the high mountains to commune with the stars and when we scour both deserts and icy wastes for meteoritic relics of our solar system. But it is still our experience on this particular outpost of life that will prepare us to search beyond it.

Each of these stories offers complementary ideas for how we can search for life in the universe. Each of these stories can be told as we explore our wonderful planet. It's time to step outside and discover what lessons the pale blue data point has to teach us.

2

TWENTY THOUSAND PINGS UNDER THE SEA

IN SEARCH OF ALIEN OCEANS

It is the beginning of the afternoon watch on Friday, June 16, 2017, and the exploration vessel *Nautilus* is keeping station 260 kilometers off the west coast of Vancouver Island in the northeast Pacific Ocean. Though it remains daylight outside, I am standing watch in the darkened control room located on the upper deck of the ship. Over two kilometers below me in the ocean depths, the remotely operated vehicles (ROVs) *Argus* and *Hercules* are diving through the impenetrable darkness toward the seafloor. Beside me sit the ROV pilots and navigators, their rapt faces limned in the glow of giant high-definition monitors. The dimly lit control van is tense with readiness, and the silence is punctuated by the steady exchange of navigation and depth calls as we approach our goal.

Staring at the screens in front of us, the exploration team is gripped by a scene enacted from a science-fiction movie: An alien landscape of sculpted and forbidding towers emerges from the dark abyss. Black, mineral-rich clouds are billowing from a panorama of irregular chimneys that tower above a seemingly industrial landscape of fractures and fissures in the Earth's surface. Some of these towers are giants, such as the aptly named Godzilla, which grew to a height of forty-five meters before collapsing under its own colossal weight. Other chimneys, smaller and more delicate, display fluted branches and fans, each an underwater candelabrum decorating a landscape of gothic fantasy. Clustered around these smoking giants lie a multitude of diminutive vents, each belching away like the exhaust of a miniature engine, while others, emitting cooler, clearer fluid, create mirages of shimmering water in a scorched and blasted desert.

We have come to study the living organisms that call this alien world their home. It is a world devoid of sunlight where life at its most basic—that of bacteria and archaea—is powered by geological heat and chemistry. From one perspective this is a deeply primordial location where one may glimpse the still-smoking fires of the ancient furnaces that have powered planet Earth since its formation. Yet the scene unfolding on the screens before me also reveals a possible future where one can imagine the descendants of present-day ROV technology exploring the vast ocean worlds of our solar system and where, on moons such as Jupiter's Europa and Saturn's Enceladus, we may one day look for new and truly alien life.

Arriving at Jupiter in 1995, NASA's *Galileo* probe began the first detailed exploration of our solar system's largest planet and its retinue of moons. One of *Galileo*'s first discoveries was also an unexpected one. The spacecraft was equipped with a sensitive magnetometer that was originally intended to study Jupiter's magnetic field—the strongest among the planets of the solar system. Yet it also discovered a weak magnetic field emanating from Jupiter's second-closest moon, Europa. Oddly, this field rotated not once every 3.6 days, as Europa does, but every 9.8 hours, in step with Jupiter itself. The *Galileo* science team realized that Europa's magnetic field was induced by Jupiter's—an electromagnetic echo from its parent planet. It rapidly became clear to the astonished scientists that to sustain that echo Europa would require an electrically conductive layer beneath the surface ice sheet. Although it was masked in the abstract language of electromagnetism, the science team had just discovered the largest ocean anywhere in the solar system. The *Galileo* observations revealed a lunarwide ocean on Europa, up to 100 kilometers deep and containing a volume of liquid water equal to two times that of Earth's own oceans. Internal heat, caused by rhythmic tidal forces raised as it orbits Jupiter, maintains Europa's ocean in a liquid state, and the presence of salt—inferred from the same magnetic field measurements—suggests that liquid water is in direct contact with the rocky core of the moon.

A different spacecraft orbiting a different planet would tell the same story of an ocean-bearing, icy moon. In 2005 NASA's *Cassini* probe in orbit about Saturn would perform a close flyby of its satellite Enceladus. This tiny moon is only some 500 kilometers across. That's six times smaller than Europa, and, if you could find a highway, it would take the same driving time as travel between Pittsburgh and New York City (think Paris and Amsterdam if you are reading this in Europe). *Cassini*'s onboard camera showed plumes of ice crystals venting from giant fissures in the south pole of Enceladus's icy outer crust. Later measurements would confirm this to be the result of salty, liquid water escaping from a subsurface ocean. The astonishing—and unconfirmed—implication is that if an ice-bound moon this small can host an inner ocean, then what of the fourteen larger moons in orbit about the outer planets of the solar system? Might these distant and ice-bound moons provide astrobiologists with a chance to ask whether life arose independently not once but multiple times in the outer solar system? Though this is a deeply compelling idea, we risk sailing into speculative waters, and perhaps we would be wise to limit ourselves to searching for life on just one moon first.

It was these geological clues uncovered by *Galileo* at Europa and *Cassini* at Enceladus that had drawn me to join the *Nautilus* and visit Endeavour, one of planet Earth's most dramatic deep-ocean hydrothermal vent systems. Yet to understand the trail of discovery that has led scientists to contemplate the astrobiological connection between hydrothermal vents and icy solar system moons, we must travel back through some sixty years of scientific history to revisit the story of a modern paradigm shift in how we viewed our own planet.

We like to think that we know our home planet fairly well, yet it is humbling to learn just how recently much of that knowledge was revealed—and to contemplate what further discoveries await us. For to tell the story of how we discovered deep ocean hydrothermal vents and their connection with the ocean-bearing moons of the outer solar system, we must first learn how we came to discover planet Earth itself. At the beginning of the 1960s the depths of Earth's oceans were as unknown as the surfaces of the planets in the solar system. They represented a truly alien world.

Perhaps that is why our oceans fascinate me to the degree they do as, over a ten-year span through the tumultuous and wonderful 1960s, scientists explored this alien world under the waves and thus came to understand the detailed inner workings of our own planet as a result.

Prior to this age of discovery, maps of the ocean floor stretched smooth and featureless across the gulfs that separated the continents. In 1912 Alfred Wegener, a German meteorologist, had pointed out the curious fact that cutout shapes of the Earth's continents could be made to fit together to form a planet-sized jigsaw puzzle. However, his accompanying theory of "continental drift" was snubbed by the geological establishment because it lacked a mechanism to propel entire continents across the surface of the globe. Then, in 1931, the British geologist Arthur Holmes proposed the idea that giant convection cells in Earth's mantle might act to transport heat from Earth's radioactive core to the planet's surface, and that this in turn could provide the motive force required by continental drift. Yet the critical observations needed to confirm these ideas were not yet in place, and the scientific community—inert as all scientific communities are wont to be—was slow to shift the weight of its opinions.

Beginning in the late 1950s, however, geologists and oceanographers began to map the ocean depths using the forerunners of today's sonar devices. A compressed-air gun, towed behind a survey ship, would be discharged to create a sudden blast of sound waves that, when reflected from the seafloor, could be used to compute the depth beneath the ship from the intervening delay in the echo. Each ping represented a datum of knowledge, another depth sounding on a previously featureless chart. As oceanographic pioneers such as Bruce Heezen, Marie Tharp, and Maurice Ewing filled in the blank spaces in a series of perspective-shifting physiographic maps, there came the realization that vast forces must be at work on and beneath the ocean floors. Greatest among these discoveries was an unbroken undersea mountain chain that was eventually found to straddle the world. Akin to the line of stitching running around a baseball, the mid-ocean ridge system appeared to represent the very seam of Earth itself. In addition to this discovery, a novel analysis of the location and depth of earthquakes revealed that many were generated at these mid-ocean ridges. Earth was rumbling and creaking—but was it in motion? Part of the astonishment associated with this breathless decade of discovery came from the manner in which new observations suddenly found a place in the growing pattern of results—the moment when the image on the jigsaw puzzle takes shape and the remaining pieces snap into position. It was the seemingly unrelated field of magnetic reversals that would offer the next key piece in the puzzle.

For reasons we still do not understand, Earth's entire magnetic field flips its poles during random events that have occurred at least 183 times in the past 83 million years. (Presumably for more of Earth's history as well, but these are simply the rocks for which good data is available.) The discovery of such geomagnetic reversals was itself a moment of profound—and still puzzling—insight. Yet, for our understanding of the seafloor, the revelatory observation was that the pattern of magnetic reversals recorded in rocks laid down on either side of the great mid-ocean ridges appeared to be identical. Once these magnetic reversals were placed within a time-ordered sequence, it became clear that new ocean crust was being created at the mid-ocean ridges, welling up through cracks in Earth's seam and flowing outward to create this symmetrical pattern. To accommodate the new rock, entire ocean basins moved as on a conveyor belt, away from either side of the ridges to the surrounding rim of continents. Suddenly the theory of Earth's oceans became a theory of the planet itself—plate tectonics—as we came to understand the coordinated global motion of Earth's continental and oceanic tectonic plates.

Deep ocean hydrothermal vents were one of the last pieces of the geological puzzle to drop into place. Compared with Earth's hot interior, the average flow of heat through the ocean floor was smaller than expected—a thermodynamic-shaped hole in our tectonic jigsaw. The search was on for the elusive hot spots on the ocean floor, and mid-ocean ridges provided the prime suspect. It was at a section of the mid-ocean ridge at the Galápagos rift in 1977 that the research vessel *Knorr* was the first to measure a distinctive temperature anomaly, at a seafloor depth of 2,500 meters, with its towed array ANGUS (Acoustically Navigated Geophysical Underwater System). The anomaly was small: In the space of a few tens of seconds the temperature measurements jumped from the near constant 2°C that characterizes the deep ocean to just 8°C—a small but critically significant spike. When ANGUS's onboard camera was brought to the surface for processing, it revealed a relatively benign landscape of warmwater seeps in a field of geologically young pillow lava deposits. Clustered at the base of these shimmering seeps were fields of bright white clams—as much as a foot in length—that offered the first hint of the biological wonders that would later be found at almost all deep ocean hydrothermal vents.

The discovery of hydrothermal vents solved the remaining puzzles associated with the flow of heat through the surface of the Earth. Heat does not flow smoothly everywhere through the seafloor, and hydrothermal vents, located close to the axis of mid-ocean ridges, act to short-circuit the flow of both heat and key chemical elements through the Earth's crust. The last pieces of Earth's puzzle were clicking into place. Yet once the idea of a lunar ocean on Europa had entered our scientific imagination, the question became one of whether hydrothermal vents and the geochemical pathways to life that they provide were purely a terrestrial phenomenon or if instead we had been offered an unexpected glimpse of a possible template for an alien ecosystem. It is this question—still unanswered—that has inspired a new generation of solar system oceanographers in their quest for new discoveries.

Despite the wonders hidden in their depths, our oceans ultimately represent the merest condensation of water on the surface of what remains a rocky world. We are often told that Earth is an ocean world, with over two-thirds of its surface covered with water. Yet it is worth considering the relative contribution of Earth's oceans to the overall scale of our planet and asking whether we truly qualify as an ocean world compared to moons such as Europa. The average depth of Earth's oceans is just under four kilometers. However, were you to smooth out the continents and ocean floors, leveling the mountains and filling the trenches that rise and fall across Earth's exterior, that water would flow out across the surface to a depth of about 2.5 kilometers. A world-circling ocean over two kilometers deep would certainly be profound compared to any human scale, yet its depth remains tiny compared to Earth's radius of 6,370 kilometers. A useful visualization of the disparity is to imagine leaving a basketball out in your garden overnight. In the morning it may well be covered by a thin layer of dew, perhaps to a thickness of a millimeter or so. Yet this thin layer of dew on the basketball is about twenty times as deep as our own oceans when compared with the relative size of the Earth. If you want a more dramatic comparison, consider that if we scaled the size of Europa's rocky interior to match the size of Earth, then Europa's ocean would stretch from the surface of the Earth to lap at the orbit of the International Space Station, some 400 kilometers above us.

This relative contribution of ocean depth to rocky interior is why we can justifiably call Europa an ocean world—perhaps *the* ocean world of the solar system. Yet despite dwarfing Earth's oceans in terms of relative scale, Europa's oceans may prove to be unexpectedly similar to Earth's in terms of the conditions we might discover there. When we imagine Europa's oceans plumbing to depths of 100 kilometers, we may well think of the crushing pressure that exists at the very deepest part of our own ocean—the Challenger Deep, just over ten kilometers below the ocean's surface. Surely we could not conceive of a craft capable of surviving what would presumably be the immense pressure at the bottom of Europa's oceans. However, we must remember that hydrostatic pressure in a planetary or lunar ocean is generated by the *weight* of overlying water, and that on Europa water weighs less than it does on Earth. By a quirk of mathematics, the force of gravity at Europa's rocky surface is approximately ten times less than at the Earth's surface—with the result that on Europa water weighs ten times less than on Earth. I think you can anticipate what is coming next. With ten times the depth yet one-tenth of the weight, the pressure at the bottom of the Europan ocean will be comparable to that at the deepest parts of Earth's oceans. At one stroke, this profound realization changed the nature of ocean exploration here on Earth. Henceforth ocean scientists and astrobiologists would study Earth's oceans in partnership, joined by the common aim of one day exploring new oceans—at once alien and familiar—out on the high seas of the solar system.

It was this trail of geological and astrobiological clues that had drawn me to the bright and blustery seas of the northeast Pacific, where the E/V (exploration vessel) *Nautilus* was to be my home for three weeks during our science cruise off Vancouver Island. I had joined the ship as a science communication fellow, one of a corps of formal and informal educators invited aboard to join the scientific team in bringing the story of modern ocean exploration to the public. The *Nautilus* is operated by the Ocean Exploration Trust, whose president and founder, Robert Ballard, is a marine geologist who has spent a lifetime exploring the world's oceans with innovative technology. Though he is perhaps best known for discovering the wreck of the *Titanic* in 1985, Ballard was also a team member aboard the R/V (research vessel) *Knorr* in 1977 and was present at the beginning of the story of hydrothermal vents.

My own voyage with the *Nautilus* was part of a scientific cruise in partnership with Ocean Networks Canada (ONC) and was named Expedition 2017—Wiring the Abyss. ONC is an initiative of the University of Victoria in British Columbia and is dedicated to operating deep-sea observatories off the Pacific, Arctic, and Atlantic coasts of Canada. Our three-week voyage would take us on a circuit around its NEPTUNE observatory—the North East Pacific Time-series Underwater Networked Experiments observatory—an 840 kilometer loop of power and fiber-optic cable that links a network of instruments on the seafloor to scientists ashore. The area of seafloor on which NEPTUNE is constructed spans one of the smallest tectonic plates on Earth, the Juan de Fuca plate—squeezed between the giant Pacific and North American plates—and the site was chosen because it offers researchers access to a wide range of deep ocean geological environments within a relatively small geographical area. The scene I described at the beginning of this chapter took place at one of the deepest and most dramatic sites encompassed by NEPTUNE. Endeavour is located at depths of 2,200 to 2,600 meters and is part of the Juan de Fuca Ridge—the boundary between the Pacific and Juan de Fuca tectonic plates. It is an active spreading center of the Earth's crust and hosts some of the most intense hydrothermal activity on the planet.

This is where the *Nautilus* comes in. Named after Captain Nemo's ship in Jules Verne's novel *Twenty Thousand Leagues Under the Sea*, this ship, sixty-four meters long, offers an explorer's home away from home for seventeen professional crew members and thirty-one scientists, engineers, and educators. Our role on the cruise would be to support the scientists of ONC in connecting new instruments to the seafloor grid, maintaining existing equipment, and retrieving long-term experiments for analysis ashore. Daily life aboard the *Nautilus* is dominated by the rhythm of dive operations. When the two ROVs, *Hercules* and *Argus*, are in the water the *Nautilus* operates for twenty-four hours a day, with each crew member standing two four-hour watches during that period. I would stand my watch from 4:00 to 8:00 a.m. and 4:00 to 8:00 p.m. each day. When not on watch, in addition to the daily round of eat, sleep, repeat, each crew member would have additional responsibilities particular to their roles. In my case, as a science communication fellow, my off-watch activities would find me in the compact and bijou broadcast studio aboard ship presenting live educational broadcasts to schools and museums across the globe.

Personal real estate aboard an oceangoing research vessel is definitely at a premium. My own haven while at sea was a bunk in cabin 81—located in the deepest, darkest section of the vessel. Such a cabin might seem a poor deal, the short straw drawn by those on the lowest rung of the crew ladder. However, I soon discovered that being close to the ship's roll axis meant that heavy seas would disturb me less than my crewmates higher up in the ship (literally and figuratively), and the lack of a porthole turned out to be a welcome blessing during my midday nap—a basic survival strategy while aboard.

The daily watch schedule offers each crew member one individually tailored trial of fatigue to undergo each day, and my own test would begin at 3:00 a.m. with the strident call of my alarm waking me from my gently rocking slumber. Keeping the lights off was a basic courtesy offered to my cabin mates as I stumbled from one brace position to the next en route to the showers to soak up five joyful minutes of blessed hot water and find rebirth as a wakeful scientist. A brief breakfast provided the opportunity to digest the night's dive reports along with a few mouthfuls of bread and jam. Where are we operating? What tasks from the dive plans have been achieved? What is waiting for us down there? The remaining minutes before we ticked over to the watch change offered a moment to enjoy the final ritual—a few sips of hot, life-giving tea—and the chance to savor the silence of the predawn deck before the precipitous climb to the control van on the topmost deck of the ship.

The control van consists of two shipping containers that have been welded together to create a close-fitting operations room. Within this space the ten-strong watch team occupies two rows of seats in front of racks of computers and monitors. The room is dominated by two giant wall-mounted HD panels that display the video feed from *Hercules* and *Argus*, and the control van is cast in a dim, penumbral light by the unflickering glow from this array of screens. In the shrouded darkness crew members switch positions with their opposite number on the previous watch, and, after the brief and snatched conversations updating us to the job at hand, we are finally seated and ready for our watch to begin.

As a science communication fellow, my job while on watch is to occupy the comms seat in the control van—acting as the human link between the watch crew and the public ashore following us on nautiluslive.org. The online audience—you among them—sees and hears exactly what the exploration team in the control van sees and hears. The experience is shared directly and immediately by a live satellite link. Should you message the team, your words will appear on the screen in front of me, and my primary role is to be your representative in the room: to ask your questions and bring your voice into the conversation in the control room. Perhaps more than any other experience I describe in this book, the opportunity to engage with the crew of the Nautilus is your chance to "get in the room" with modern scientific exploration.

As related in the opening scenes of this chapter, you are offered a ringside view as the exploration team—now just starting their watch—bring their entire focus to bear upon the deployment of *Argus* and *Hercules* at the hydrothermal vents of Endeavour. It is these two ROVs, under the command of the pilots seated next to me in the control room, that make up the final links in the chain of technology connecting both you and the exploration team to the events occurring on the seafloor. The Argus of antiquity was a giant commanded by Hera to watch over the nymph Io with his one hundred eyes, and the present-day ROV *Argus* lives up to its name by acting as a chaperone to *Hercules*. From the stern of the *Nautilus* an armored cable runs down into the dark seas. It connects to *Argus* in the depths below and provides a power and fiber-optic data link to the ROVs. However, *Argus* remains tethered to the ship and is pulled up and down in response to the tug of the Nautilus as she rides the surface swells. To isolate this rhythmic pull from its partner *Hercules* while maintaining the power and data link, the two ROVs are themselves tethered together by a fifty-meter high-visibility cable. Being neutrally buoyant, this cable floats freely in the water and acts to protect *Hercules* from the steady rise and fall of the ocean waves. Working as a team, *Hercules* and *Argus* can explore to depths of up to four kilometers. Argus provides the "eye in the sky," while Hercules operates as a completely stable work platform. With two manipulator arms, a suite of thrusters, and HD cameras, it provides its pilot aboard the *Nautilus* with a compelling and immersive sense of presence on the seafloor.

The Endeavour hydrothermal vent systems that we encountered at the start of this chapter are associated with a mini mid-ocean ridge just off the coast of Vancouver Island on the western shores of Canada. The fissure between the Pacific and Juan de Fuca tectonic plates allows magma in the liquid mantle to approach far closer to the surface than anywhere else on the planet—to within one kilometer in some locations, compared to five to ten kilometers in most other areas of the ocean crust. Seawater percolating down through faults and fissures into the seafloor is heated near the magma, and this abundance of energy powers a host of aqueous chemical reactions that saturate the superheated water with dissolved minerals—chief among them sulfides of iron, copper, and zinc.

As water is heated within the rocky depths, it becomes less dense and therefore more buoyant than the seawater above it. As a result, the faulted and fractured rocks overlying the boundary between these tectonic plates act like a complex plumbing system, with cold seawater descending and hot, mineral-rich water ascending though the rock layers. On reaching the seafloor, these plumes of water, which can be as hot as 400°C, encounter seawater with an ambient temperature of 2 to 4°C. The crushing pressure at Endeavour prevents the superheated water from becoming steam, yet contact with colder water causes the dissolved minerals it carries to suddenly and dramatically precipitate out of solution. In particular, as the iron sulfide precipitates it forms dense hazes of fine black particulates that, when viewed underwater, appear as belching clouds and give these vent systems their common name of black smokers.

The temperature gradients around the vent systems are immense and test the nerve of any submersible or ROV pilot with the temerity to explore them. Water escaping at the very base of a vent may be as hot as 400°C and rises in a turbulent, expanding plume. However, a probe one meter off to the side will still register the temperature of ambient seawater, say 3°C in this case. Even as the probe is edged closer, to within a few centimeters of the plume base, it will register a temperature of only 20°C or so. Upon their first encounter with black smokers, the pilots of the submersible *Alvin* used a manipulator arm to insert a temperature probe directly into water escaping from the base of a vent. The probe, constructed of the same acrylic glass as the viewing ports, promptly melted, and the pilots beat a measured yet deliberate retreat. Indeed, one treasured memento from my visit to Endeavour is a section of three-quarter-inch electrical cable in a thick plastic sheath. It had the misfortune to fall across the opening of a seemingly innocuous vent, and in the space of about one second it melted and parted—now appearing like an oversized stalk of bright green asparagus sprouting from my desk.

In a world without sunlight, these vents form the base of the food chain, the geochemical foundation on which the entire ecosystem rests. Therefore, upon arriving at Endeavour, one of our first tasks was to install a probe into the water emerging from a particular vent location and determine the concentration of dissolved mineral ions within the vent fluid. The device used, known as a benthic and resistivity sensor (BARS), consists of a custom-built titanium pressure cylinder housing the electronics needed to measure the temperature and resistivity of the super-heated water. About fifteen centimeters long and set on short stubby legs, the BARS receives data from a ceramic probe inserted into the base of a nearby vent system. This is one example where continuous monitoring can provide critical insight, as the temperature and chemistry of the emerging vent water can vary rapidly in response to changing conditions in the reaction zone several hundred meters below. However, for the measurement to be of value the water must be sampled as it emerges from the very base of the vent, right on top of the fissure in the volcanic rock.

I will admit to being shocked that my first encounter with a hydrothermal vent was to witness *Hercules* topple a two-meter tall tower of encrusted mineral deposits in order to expose its foundation of volcanic basalt. Had I

Black smoker with life and BARS A benthic and resistivity sensor (BARS) in its natural habitat, in this case, two kilometers under the ocean and perched on a black smoker vent at the Endeavour site. Notice the bright white tube worms (*Ridgeia piscesae*) clustered on the vent outcrop up to the very periphery of the superheated water outflow. Credit: NOAA/PMEL/EOI.

come all this way to conspire in what appeared to be an act of scientific vandalism? My misgivings largely arose from my unfamiliarity with the environment. Hydrothermal vent chimneys have been observed to grow at prodigious rates—up to thirty centimeters in a day and by as much as five meters over the course of a year. BARS devices have themselves been embedded within tall chimneys that have enveloped them during their time on the seafloor. Furthermore this relatively small chimney was a sufficiently recent feature to not yet have been colonized to any visible degree by the abundant forms of life that we observed elsewhere at Endeavour.

Since their discovery in 1977—not even one human lifetime ago—we have learned that hydrothermal vents represent transient, almost fickle, environments that can change unpredictably from one scientific visit to the next. The subsurface plumbing that underpins the towers and chimneys that we observe at the surface is anything but fixed. The hot-water supply can turn on and off. New vents can appear under carefully placed scientific equipment— with terminal results—and existing vents can suddenly be cut off from the life-producing flow of energy-rich fluids. In this case the mobile residents such as shrimp, worms, and spiders can jump ship and land on the shores of nearby vents. Meanwhile an entire generation of fixed residents such as tube worms may be left to wither and die even as their larval young, the promise of a new generation, float freely away on the weak currents. More striking still are events that may occur deep in the reaction chamber when magma retreats back into the ocean crust. When this happens, entire vent regions can cease to flow, leaving behind a landscape as stark as any ancient city lost to time: a geological relic devoid of life.

On a personal note, any chagrin I felt at my part in toppling the chimney was replaced with scientific wonder when I later was able to handle a sizable chunk of vent material returned to the surface by *Hercules*. The friable gray rock has the strange consistency of tightly compressed cigarette ash and, not unlike more familiar rocks such as pumice, is much lighter than it looks. Sadly, with time this prized sample will dry out and crumble, the particles of iron sulfide steadily rusting as they are oxidized in our atmosphere—my own transient reminder of the ephemeral nature of vent chimneys and the hydrothermal activity that creates them.

Geologically speaking, these vent systems were a revelation when they were discovered in 1977: They provided the missing link in understanding both heat flow through Earth's crust and the chemical composition of its oceans. What was completely unexpected, however, was that such environments, well below the reach of any sunlight, would host abundant and unique ecosystems.

When the research vessel *Knorr* discovered the first vent systems at the Galápagos rift, it looked out over a landscape that we today refer to as warm seeps. We now recognize that these relatively benign, lower-temperature flows occur on the periphery of rift zones, away from the main axis of activity at the mid-ocean ridge. It was a joint American-French expedition in 1979 to the site known as 21°N, located off the southern tip of Baja California, that first revealed the dramatic and alien landscapes of the black smokers— pulsing at the very heart of the active boundary between the Pacific and Rivera tectonic plates.

The first scientists to visit these black smoker vents were simultaneously astounded and unsettled by visions of abundant yet profoundly unique forms of life. Silent gardens of giant tube worms wafted in the turbulent currents at the periphery of the vents, their vivid red bodies and gills in stark contrast to their white calcareous tubes. Ghostly white crabs prowled through this forest of tubes and frond-like gills, snipping off tasty morsels from any tube worm too hesitant in withdrawing. Gauzy mats of filamentous bacteria were being harvested by hosts of eyeless shrimp, some of which carried their own portable gardens of bacteria growing on their undersides.

To explore here was a deeply unfamiliar experience, one for which the language of the surface world was lacking in appropriate imagery. In Greek mythology Hades provided his captive wife Persephone with a garden of otherworldly, ever-blooming flowers as a comfort during the six months each year she was forced to spend in his cold, sunless realm. Yet here scientists had discovered a vivid evocation of Persephone's garden, apparently sustained by the undying fires of a present-day Hadean abyss.

The abundant biology discovered at hydrothermal vents posed the fundamental question of what chemical process supported these complex ecosystem, where the pressure is as high as 260 atmospheres, there is no sunlight whatsoever, and the water temperature hovers just above freezing. Who are the primary producers that lie at the base of the food chain and support all other life in these strange new worlds? On Earth's surface, caressed by the sun, the primary producers are familiar to us—the photosynthetic plants and microbes that harvest the energy contained in solar photons. Yet here, in the aptly named Abyssal Zone, far beneath the reach of the sun, what source of metabolic energy could provide the raw material for life?

To understand how the flow of electrical energy—electrons—powers life, we need to go back to the basics of photosynthesis. The raw materials used by a plant or microbe to create a simple sugar such as glucose are carbon dioxide, water, a burst of energy, and some free electrons. Carbon dioxide and water come from the environment in which the plant or microbe lives. We call such creatures *autotrophs*—from a union of the Greek words for *self* and *feed*—as they obtain the raw materials they require dissolved in the water around them. So where do the electrons and the energy required to construct the molecule of glucose come from? In the case of photosynthesis, the magic molecule is chlorophyll, which absorbs the energy contained in solar photons and emits an energetic electron in response. Organisms that do this can be labeled further as *photoautotrophs* as they are sunlight-powered self-feeders. Chlorophyll remains electrically neutral by stripping electrons from hydrogen in water, creating protons (H^+) and molecular oxygen (O_2). For the last link in the chain we need to understand where the energy required to assemble the CO_2 and H_2O into glucose comes from. In this case it comes from the electrons themselves. Rather like the way a sailing ship captures a fraction of the energy contained in the passing wind, a host of enzymes pass the free electrons between themselves in a cyclic dance that produces energy-bearing molecules of adenosine triphosphate (ATP)—the metabolic currency of energy exchange used by all life on Earth. When converted to adenosine diphosphate, ATP releases the necessary spark of energy.

In the ocean depths, oxygen and carbon dioxide are constant constituents, dissolved in the abundant seawater. Yet where does the supply of free electrons come from in the absence of chlorophyll and the sunlight required to power it? The key resource turns out to be hydrogen sulfide (H_2S) dissolved in the vent fluid. Though electrically neutral, this rotten-smelling gas exerts only a weak grip on its outer shell of electrons. Oxygen, on the other hand, is the avid seeker of electrons—always on the look out to add another prize to its collection. It is the transfer of charge from the electron donor (hydrogen sulfide) to the electron acceptor (oxygen) that sets in motion the very same cyclic dance of electrons that powers the creation of ATP in photosynthesis—the Calvin cycle.

Yet this is not photosynthesis but *chemosynthesis*—the production of simple sugars using chemical rather than solar energy. The shift from *photo* to *chemo* is subtle yet profound. At one stroke the change of prefix liberates life at hydrothermal vents from any dependence upon sunlight. To astrobiologists, always alert for new potential habitats in the solar system, the discovery that life could exist, even thrive, in the somber recesses of Earth's oceans immediately offered the prospect that alien life, powered by a similar source of geochemical energy, could thrive in the dark, ice-capped oceans of Europa and Enceladus. On Earth the organisms that tap into this source of energy are known as *chemolithoautotrophs* (adding *litho* acknowledges the fact that inorganic minerals are the source of the chemical energy). Hydrogen sulfide, carbon dioxide, and oxygen provide an abundant and constantly renewed source of geochemical energy for colonies of microbes. However, what was poorly understood at the time and has since been revealed in a number of elegant studies is the extent of the symbiotic relationships between the macrofauna and microbes at hydrothermal vent sites.

The first scientists to peer through the viewing port of the submersible *Alvin* were confronted with a pressing question: What were all of these abundant and highly visible organisms—the macrofauna—feeding on? In many ways these first explorers of hydrothermal vents were astrobiologists as much as they were biologists. True, the presence of clams and tube worms confirmed that biology was indeed present, but a critical piece of the puzzle was missing. In this case it was the uncertain presence and nature of the microfauna—microbes—underpinning the visible life. In this context the wonderful 1979 paper "Chemosynthetic

Primary Production at East Pacific Sea Floor Spreading Centers," by marine microbiologists Holger Jannasch and Carl Wirsen, represents a prescient template of what might one day confront an astrobiologist exploring hydrothermal vents on Europa. At the start of our journey together I invited you to consider several practical definitions of life—order, metabolism, and evolution—all three of which would eventually be applied to understand the microbiology of hydrothermal vents.

Jannasch and Wirsen began with the search for cellular order in the milky vent fluids at the warm seeps they were studying. They pumped these turbid fluids through a filter, and when later placed under a powerful microscope they were able to observe a multitude of individual microbial cells in various stages of growth and division. Using our high school grading analogy, Jannasch and Wirsen would have scored an initial grade of 33/100 for this result. The hunt for metabolic activity came next, and the authors performed a test that carried with it echoes of the *Viking* labeled release experiment undertaken on Mars only three years earlier. A sampling arm on *Alvin* was used to fill six syringes with vent fluid, each syringe having been dosed with radioactively labeled CO_2. Whatever was in the vent fluid then actively took up the labeled carbon from the CO_2 over the next ten days—a result consistent with metabolic action. If we recall that, in contrast to the Martian soil sample studied by *Viking*, the vent fluid contained no oxidizing chemicals that could mimic this result, then our score after this second test would be 67/100. Finally came "I evolve, therefore I am." Jannasch and Wirsen correctly foresaw that their results would stimulate much more detailed follow-up work to determine explicitly which species of microbes might be present at hydrothermal vents. Later teams of scientists created cultures of microbial samples that had been filtered from vent water and their periphery. The science of genetic sequencing was at the time undergoing its own technical revolution, and the small yet distinctive genetic fingerprint carried in every living creature's 16S ribosomal RNA molecule was used to construct the complex genetic map of hydrothermal vent microbes. Three distinct tests for life and three clear successes gives a score of 100 percent. Although this discovery represented a success for biology, not astrobiology, the methods used to confirm the existence of chemosynthetic microbes at deep ocean hydrothermal vents provide a compelling primer for any scientist who contemplates sampling the waters of Europa or flying through the geysers of Enceladus. We could easily get lost down the rabbit hole of all the wonderful biological relationships that researchers have discovered at hydrothermal vents. Yet I would be remiss as a guide if I failed to highlight just a couple of what are, to me, the most compelling examples of evolutionary adaptation and cooperation in these environments. I do not claim that these examples offer a glimpse behind the curtain that currently obscures our view of what life might look like on Europa. However, while we are here, it would be silly not to take a peek . . .

Giant tube worms are among the most visibly striking inhabitants at deep ocean hydrothermal vents. Two of the most common species are *Riftia pachyptila*—the larger of the two, found at vent sites throughout the Pacific basin—and *Ridgeia piscesae*—a more diminutive species that dominates at particular sites such as Endeavour. Despite their varying dimensions, both *Riftia* and *Ridgeia* are vestimentiferan worms—the muscled walls of the vestimentum (a section of the trunk) allow them to glide up and down within their white calcareous tubes to extend their blood-red gills into the mineral-rich vent flows. When first discovered, *Riftia* posed a crude anatomical conundrum to the scientists who studied it. They were quite literally found to be "gutless and buttless"—lacking not just an intestinal tract and anus but mouthparts as well. They appeared to be flourishing—growing to over two meters in length—while flouting the normal rules of "what goes in must come out." Despite this apparent lack of feeding apparatus, the internal spaces of *Riftia* and *Ridgeia* are dominated by an organ known as the trophosome, which can make up as much as three-quarters of their body volume.

Many of the ecosystems we have encountered at hydrothermal vents appear strangely inverted, even fantastical, and perhaps none is more so than the trophosome—an internal hanging garden where the microscopic fruits of boundless chemosynthetic microbes hang suspended within the tube worm's elongated body. A healthy fourteen-gram tube worm may contain up to 40 billion bacterial cells within its trophosome—to put that number in perspective, you or I might poop out a similar number of methanogenic gut bacteria each day. Each respiring *Riftia* carefully nurtures its crop with a steady suffusion of hydrogen sulfide, oxygen, and carbon dioxide that the chemosynthetic bacteria metabolize into simple sugars—food for the tube worm—and microscopic crystals of elemental sulfur. It is a symbiotic relationship that appears to be consummated at birth, as larval tube worms floating in the water column retain their ancestral mouths and digestive tracts and are thought at this stage of their lives to ingest the bacteria that will seed their later harvests.

The second example of vent life that I can't pass by is *Rimicaris exoculata*—the ridge-dwelling shrimp deprived of eyes—found at hydrothermal vents at the mid-Atlantic ridge. They survive by farming the mats of filamentous microbes that populate the warm hinterlands of black smoker vents, in addition to hosting portable gardens of microbes on the underside of their shells. These blind cultivators of chemosynthetic bacteria deserve a place in our story thanks to a seemingly cruel evolutionary regression that may yet link them to life far beyond the solar system. Though devoid of the relatively complex eyes characteristic of the shrimp family, *Rimicaris* sports a double patch of the red-light sensitive pigment rhodopsin beneath a transparent cuticle in its carapace. You might consider that the absence of eyes represents a worthwhile evolutionary cost-cutting exercise in the lightless depths, yet that scenario is not completely accurate.

Were you to view a hydrothermal vent through the sensitive optics of a thermal imaging camera, you would see them shine like so many lighthouses in the abyssal darkness. In this case the light is produced by the thermal motion of the hot molecules of water, and when viewed against the backdrop of the colder ocean and seafloor the vents shine like radiant beacons. The red patches on the backs of *Rimicaris* provide these shrimp with sufficient sensitivity to the thermal photons emitted by the vent fluid to safely (at least most of the time) navigate the perilous periphery where vent fluids mix with ambient seawater.

The physical process that produces the infrared radiation emitted by the vent fluids is the same one that generates visible photons from the hot photosphere of the sun. The only difference between the two is temperature. The vent fluid is cooler than the surface of the sun and therefore emits fewer photons that display on-average redder wavelengths. Yet it is clear that evolution has provided an ocular adaptation that allows *Rimicaris* to benefit from this radiation as a navigation tool.

It is interesting to note that the radiation produced by hydrothermal vents occurs at infrared wavelengths similar to those produced by cool, dim red dwarf stars such as Proxima Centauri, the closest star to our sun and one that hosts its own Earth-sized rocky planet. Are there as yet undiscovered terrestrial microbes at hydrothermal vents that have adapted the biochemistry of photosynthesis to exploit this abundant supply of infrared photons in the otherwise black abyss? Might such extreme terrestrial microbes offer an unexpected vision of life beyond the solar system? It is a far-out idea that remains speculation—the typical photon energy emitted from Promixa Centauri is five times greater than that emerging from a hydrothermal vent. However, the potential it offers us to imagine a truly alien photosynthesis operating under the feeble glow of a dim and distant sun makes me reluctant to dismiss it without further exploration of the wonders exhibited by hydrothermal vents.

It may sound trite to repeat the formula that with the discovery of hydrothermal vents an alien world was revealed before our eyes. It is more correct to acknowledge that what was actually being discovered were hitherto unthought of and wonderful manifestations of terrestrial life. Yet parallels remain to be drawn, arcs of scientific inference that link what we continue to learn at terrestrial deep ocean hydrothermal vents with the possibility of life existing in the yet undiscovered depths of oceans on worlds beyond Earth. It was a vision that was apparent from the very beginning: On seeing these ecosystems for the first time in

1979, the late Holger Jannasch made the immediate astrobiological connection. "We were struck by the thought, and its fundamental implications," he once recalled, "that here solar energy, which is so prevalent in running life on our planet, appears to be largely replaced by terrestrial energy—chemolithoautotrophic bacteria taking over the role of green plants. This was a powerful new concept and, in my mind, one of the major biological discoveries of the 20th century."

Though I have presented the evidence for a lunar ocean on Europa in terms of the *Galileo* magnetometer results obtained in 1995 (as they represent what is currently definitive), it is important to note that when *Voyager 2* flew by Jupiter in 1979 it provided our first close-up view of Europa. The images received, byte by byte, across the Deep Space Network were reassembled to reveal a fractured, icy surface that was young enough to be devoid of impact craters. These images in turn sparked a flurry of academic papers speculating on the role that tidal heating by Jupiter might play in warming the lunar interior and offering the first possibility of an ocean world beyond Earth. At the same time that hydrothermal vents and their associated biological wonders were impacting our consciousness, we had been offered a scientific counterpoint, a second voice in a cosmic fugue that, when heard together, sang a story of life that could flourish in the dark waters of a distant moon.

I find such moments of scientific synchronicity deeply fascinating—especially since we often view them through the partial lens of history. What must it have been like to be at a scientific conference in 1979? To be an oceanographer or a planetary scientist? Were these bursts of scientific revelation akin to being told of long-held family secrets? Was it like discovering, as an oceanographer, that your distant kinfolk in planetary science were instead among your closest relations—an unknown, unguessed-at scientific sibling? Despite this profound connection between terrestrial and astrobiological ideas, there remains one important caveat that we cannot avoid. The deep-ocean, sulfide-oxidizing bacteria and archaea that I have described so far remain metabolically reliant on surface photosynthetic organisms (plants, algae, and bacteria) that provide all of the dissolved oxygen in Earth's oceans. Anaerobic chemosynthetic bacteria and archaea—microbes that can metabolize geochemical energy without using molecular oxygen—are present but are rarer and less well studied than their oxygen-breathing brethren. Their rarity compared to the much more abundant aerobic microbes could well be due to the unique metabolic energy advantage offered by using oxygen as an electron acceptor rather than any other atom or molecule. As far as anaerobic microbes are concerned, in the race to exist their opponents are quite literally turbocharged. Do such anaerobic microbes represent ancient relics, the original inhabitants of hydrothermal vents that were usurped by aerobes—oxygen burners—once dissolved oxygen accumulated in the oceans some time before the beginning of the Proterozoic eon 2.5 billion years ago? The key assertion of the idea of the pale blue data point is that the study of astrobiologically relevant locations on Earth can inform the search for life beyond it. Is the life we observe at vents today truly primordial? This is a question that haunts the minds of those who claim that deep ocean hydrothermal vents provide a possible view of how life began on Earth. Furthermore, do present-day hydrothermal vents offer a good model for life in a Europan ocean where surface photosynthesis has not suffused the waters with dissolved oxygen? Clearly there is still much to learn.

All these worlds are yours. Except Europa. Attempt no landing there. So we are told by the monolithic aliens in Arthur C. Clarke's *2010: Odyssey Two*, who seek to protect the nascent life forms nurtured in the oceans beneath Europa's icy surface from human interference. Yet, having explored the depths of Earth's oceans in search of Europan analogues, a scientific return to Europa is exactly what we must now contemplate. Clarke's Odyssey series is fascinatingly prescient in so many ways, and I admired its themes so much that I paid homage to the "all these worlds" riff by shamelessly taking it as the title of my first book. In *All These Worlds Are Yours* I highlighted the potential of sending an orbiting spacecraft to the tiny yet ocean-bearing ice moon Enceladus as a practical life-detection mission. The probe could then dive through the great geysers of salty water vapor and ice crystals erupting into space from the subsurface ocean and return that material to Earth for analysis. Such a prospect remains enticing and lingers hopefully on the periphery of NASA's drawing board of future missions. However, this time around I wanted to dive more deeply—even though that means being more speculative. What would a flagship mission to Europa look like? What known engineering challenges would have to be surmounted, and where would we confront the cloaking darkness of the unknowns concerning such an ambitious undertaking?

NASA's near-term plans are centered on *Europa Clipper*, a Jupiter-orbiting spacecraft that will perform multiple close flybys of Europa itself. *Clipper* launched in 2024 atop a SpaceX Falcon Heavy rocket and will take just over six years to tread the cold and silent road to Jupiter. Of the many scientific instruments carried aboard, I am most excited about REASON—the Radar for Europa Assessment and Sounding: Ocean to Near-surface. REASON will reveal the interior structure of the Europan ice sheet, using proven technology that has mapped the inner structure of both the Antarctic ice sheet on Earth and the polar caps of Mars.

Europa Clipper and REASON will create a 3D map of the ice sheet and will reveal some, though not all, of the secrets that it contains. The arcs traced by *Clipper*'s close flybys will allow it to study only a small portion of Europa's surface. However, given that we currently don't know how thick the ice even is, a partial map of Europa's interior will signify a marked increase in our knowledge. The surface images we have from the *Galileo* probe show smooth plains of ice punctuated by great rifts and fractures. Chaotic terrain appears to have been fractured asunder and joined back together with flows of more mobile ice. What lies beneath this enigmatic surface? Do surface rents plunge through to the ocean below? Does liquid water rise through the ice sheet to approach the surface just as magma and lava do through Earth's exterior crust? For astrobiologists the important question to be answered is, How close does liquid water approach the surface of Europa?

Clipper's 3D map of the Europan ice sheet will provide an invaluable chart that will guide the next generation of explorers who seek to place a lander on the presently unknown surface. For all that can be revealed from an orbital flyby, it is a lander that will deliver the ultimate ground truth. Samples of surface ice and all that it might contain would be excavated from a few all-important centimeters below the radiation-baked—and therefore chemically altered—surface. Detailed analysis performed by small onboard laboratories would probe the samples with new and innovative tests for life, based upon all three aspects of life that we have encountered before: order, metabolism, and evolution.

The mission planners at NASA's Jet Propulsion Laboratory (JPL) who are developing the Europa lander mission concept are focused on answering some very specific questions. What regions on the surface might have been exposed to ocean material, possibly venting from Enceladus-like geysers? Might young ice derived from buoyant pockets of liquid water in the ice sheet have erupted at the Europan surface in a manner akin to volcanic lava flows on Earth? Yet for all these questions of deep scientific import, perhaps the most pressing is one that vexes the engineers whose job it is to safely land a probe on the surface.

NASA's as yet unnamed, and unfunded, lander concept lies at the very edge of the agency's current scientific vision for Europa. Spacecraft have successfully landed on five bodies in the solar system—Venus, Mars, the moon, comet Churyumov-Gerasimenko, and Titan—and the proposed Europa lander will combine proven technology with novel engineering ideas to navigate a host of risks en route to adding a sixth world to the roll of honor.

There is no atmosphere on Europa, so there will be no aeroshell or parachute to slow the descent as was used at Venus, Mars, and Titan. Instead, the probe would execute a de-orbit burn around Europa and make a powered rocket descent to the surface. In doing so the lander would enter the same uncharted territory as the venerable *Viking* landers that touched down on the surface of Mars in 1976. Prior to landing on Mars, NASA lacked images of the surface that could reveal any feature the size of the lander or smaller. In that sense their attempt to land represented a roll of the dice, a leap of faith, hoping that blind chance would save the lander from an unlucky event such as being tipped on its side by a large bolder or disappearing over the lip of a crater. As it happened, *Viking 1* in particular was lucky—very lucky—in landing unscathed in a region strewn with spacecraft-sized boulders. The risk is a known one, and the mission planners will either have to decide to wait until *Europa Clipper* can provide a detailed reconnaissance of the icy surface before proceeding or judiciously equip their lander with the right tools to make its own decisions upon arrival.

To this end a Europa lander may deploy an onboard LIDAR (LIght Detection And Ranging) imaging camera during the descent phase to rapidly create a high-resolution 3D surface map of the landing area, which it will process in order to decide autonomously on the best spot for touchdown. The whole process of taking the image and processing the resulting map would take just under three seconds, and from an altitude of five kilometers or less the system can detect features as small as five centimeters across within a one-hundred-meter diameter touchdown bull's-eye. NASA has previously used autonomous descent procedures to guide the *Perseverance* rover to the surface of Mars in 2021, yet the uncertain terrain of Europa may offer a final challenge to a successful landing. To avoid the fate suffered by the European Space Agency's *Philae* lander in 2014—which found itself tilted precariously when it finally came to rest on the low-gravity surface of comet Churyumov–Gerasimenko—the Europa lander will attempt to land as gently as possible beneath a rocket-powered sky crane before deploying a set of flexible legs to achieve a stable and level position.

Dare we speculate further and imagine a future mission that will brave the perils of the ice sheet and explore the oceans of Europa? An ambitious ocean explorer will have to replicate the many feats performed by a Europa lander. In addition, this imaginary probe will have to boldly go much farther than any probe before it—to melt through the ice sheet and then navigate autonomously through an alien, uncharted ocean. Can we stretch our minds to reach beyond the limits of what we can currently achieve within Earth's oceans and go full James Cameron to imagine the exploration of the oceans of Europa? These are deep seas populated by equally profound unknowns; yet if we are serious in our search for life beyond Earth, they are depths we must one day venture to explore.

On Earth we have explored subglacial lakes in Antarctica by melting our way down to them with a sterilized hotwater "drill." It is an energy-hungry process, with a great stockpile of fuel required to heat the water. Would it be practical for a Europa lander to melt its way through the ice sheet? This is speculative stuff, but in a spirit of simplicity and practicality, the ice could be melted with the passive heat produced by the kind of radio-isotope thermoelectric generator used to power the current generation of deep space probes. This idea has two important advantages: Although such generators produce only enough electrical energy to power a handful of light bulbs, they produce a similar amount of "waste" heat as a toaster oven, and they release that heat over a span of decades. The drawback, of course, is that on a moon such as Europa you may have ten kilometers or more of ice to melt through, and the raw physics of heat versus ice means that it will take the probe some hundred years to complete the job.

If that seems like an implausibly long time, just imagine trying to melt your way through the Antarctic ice sheet with the same amount of heat produced by an electric kettle or toaster oven. You could certainly improve on this crude estimate, shaving off a year here or there with clever engineering tricks. However, it's worth pausing to consider that the probe would trail ten kilometers of cable behind it as it descended through the ice sheet and assess whether the shifting dynamics of the ice would ultimately break that vital but fragile link to the surface. What this exercise really reveals is that I am not clever enough to think of a better route to Europa's ocean. However, if looking back on seventy years and counting of innovation in space exploration has taught me one thing, it is that one day someone else will.

Like Jules Verne's explorers in *Journey to the Center of the Earth*, will a Europa lander have to wait until the ice sheet is radar mapped to find a deep cavern or trench that might offer an easier route through to the core? Or could we equip a robot to explore these new ice worlds in the spirit of the old? Roald Amundsen, the Norwegian explorer whose pioneering journey to the South Pole traversed a previously unknown route through the Trans-Antarctic mountain chain, used a combination of well-honed arctic skills, on-the-spot decision-making, and deep reserves of endurance to achieve his goals. This model of autonomous exploration might offer the engineers who are designing and constructing the next generation of outer solar system robotic probes a practical route to leapfrog the need for systematic reconnaissance prior to a deep diving exploration mission.

This is the idea behind *Orpheus*, a novel autonomous underwater vehicle (AUV) specifically designed to explore the new Hades of Earth's deepest oceans. Although it was developed to explore terrestrial oceans, the robot's design team, located at NASA's JPL and Woods Hole Oceanographic

Institution, clearly has one eye firmly fixed on the long view to the ice moons of Jupiter and Saturn, where autonomy—the ability to explore beyond the range of human input—will be the key operational requirement. What distinguishes *Orpheus* from the previous generation of exploratory AUVs is not only its ability to map any route across the deep ocean floor using a combination of visual mapping and internal odometry (think of a deep ocean Fitbit) but its ability to share that information with sibling probes in a tag-team approach to autonomous deep ocean exploration.

Explorers such as *Orpheus* may one day explore the icy ocean moons of Europa and Enceladus free from any surface tether. Such a mission concept would hark back to the golden age of polar exploration when a team of rugged explorers would pass beyond the gaze of humanity into a silent, unknown realm, for months or years at a time, to eventually return with stories of hardship and discovery. What a wonderful pledge of trust to place in an autonomous probe—with what keen anticipation would we await news of its return. For return it must to that tenuous surface link to relay news of its travels through data uploads.

I certainly cannot claim that ocean exploration of worlds such as Europa or Enceladus is a done deal, that the challenges are all mapped out and we just require the funding to make it happen. The icy moons of the outer solar system present a profound exploration challenge, and the best I can do is sketch out the work now being done to meet some of the unknowns we may face en route. But nor can we ignore the fact that direct exploration of the oceans of Europa and Enceladus stand as the logical answer to the astrobiological questions that we are currently asking of these environments. The trail of clues that began at the site of deep ocean hydrothermal vents on Earth now stretches across the solar system, and I for one cannot wait to see where it ends.

Earth's oceans are responsible for the blue in our pale blue data point, and as we come to the end of our ocean journeys, it seems appropriate to consider the future of some of the deep ocean hydrothermal environments we have encountered thus far. Homer famously described the wine-dark waters of *The Odyssey* as *pontos atrygetos*—the unharvestable sea. The translation of *atrygetos* is fluid and uncertain. Some have used barren, others wild or restless. Although poetic license allows a certain degree of freedom, the seas of antiquity were indeed alien and, at the same time, inexhaustible—a realm beyond the ken or scope of humanity.

However, we have learned that the modern oceans of planet Earth are in fact all too harvestable. In this case it is the rich veins of minerals precipitated by deep ocean hydrothermal vents that have created a pressing need to both understand and protect the environments that I have written about in this chapter. Hydrothermal vents provide highly profitable concentrations of metal deposits—copper, iron, and zinc in addition to a host of rarer Earth metals—and while the seafloor is a technically demanding location in which to operate, in our present age such challenges are readily surmountable.

The continental shelf lying at the fringes of the deep ocean has been exploited for over a hundred years in search of oil and gas deposits. In contrast to this, the deep ocean floor, defined by the oceanic crust it rests above, is at present protected only by the fact that it remains more profitable to continue to extract land-based mineral deposits than to harvest the riches beneath the seas. But the barrier of economics may not exist forever. The first exploration and extraction licenses have been issued to prospect for minerals at deep ocean hydrothermal vents such as Solwara 1 in Papua New Guinea. This particular mining venture ended in 2019 amid financial failure, perhaps fortunately, since it extended the narrow window of time in which science can not just explore new deep ocean sites of interest but, more importantly, stay in front of the debate over what effects deep ocean mining will have on the novel ecosystems that we have explored in this chapter. But it remains a slender lead, one that is maintained by the continued efforts of a dedicated generation of scientists and environmentalists.

There are glimmers of hope—the Endeavour vent systems were designated a marine protected area by the Canadian government in 2003. Yet many other sites of international importance remain unprotected. Ultimately, the tide of demand for such minerals will either rise or fall in response to the economic forces exerted by our transition to a society based on renewable electric resources and the rare metals required to construct them. Taking a holistic view of our planet, this is a laudable aim that may offer a path to a future less reliant upon carbon and less vulnerable to the consequences of its unrestricted use. Yet one can only caution that we tread carefully on that path, lest we inadvertently trample the rare creatures and flowers in Persephone's perpetually dark gardens.

My time at Endeavour is nearly done, and my personal story that began at the start of this chapter is drawing to a close. The four hours of my watch have passed in a shimmering blur, and by the time I come back for my next shift the exploration team will have moved on to the next task of the voyage. While this particular dive lasted a total of nine hours, more demanding, complex dives with *Hercules* and *Argus* can last up to seventy-two hours. In using remotely operated yet highly capable vehicles controlled over a near instantaneous link, each dive has demonstrated the feasibility of techniques that may one day be used to explore oceans beneath the ice sheets of Europa and other moons in our solar system.

It is a deeply compelling vision that, like all futures that face us as astrobiologists, will require a combination of collective will and deep pockets to achieve. Yet I remain hopeful. In the spirit of our pale blue data point I came here to learn about these unique and breathtaking environments, and my visit to Endeavour did not disappoint. And, in parting, I treasure the fleeting glimpses I was afforded from my small corner of a dim control room of that distant future where Europa fills the screen and the words of Arthur C. Clarke haunt our thoughts.

3
SWIMMING WITH STROMATOLITES

THE HUNT FOR MARTIAN FOSSILS

"So you want to fly to an iron ore mining town in the northwest of Australia, drive two hundred kilometers into the desert to the Outback's hottest town, then follow a jeep track to a rock outcrop in the middle of nowhere, all to look at some wavy lines in rocks?"

My wife does have a habit of hitting the nail on the head. In my defense, though, the rocks in question host evidence of some of the earliest life on Earth: fossil traces of stromatolites. The term "stromatolite," when used to describe a fossil, refers to delicate layers of wavy lines in sedimentary rocks that preserve the remains of mat-like colonies of bacteria. The stromatolite fossils of northwest Australia are some of the best preserved and oldest on Earth, stretching back 3.5 billion years into Earth's history. Although getting to them is half the fun, I had little idea that the ultimate destination of my journey would be to arrive at a vision of early life on Mars.

CHAPTER THREE

If I had thought the 5:30 a.m. flight from Perth to Port Hedland would be a quiet affair, then the line of cars snaking out of the airport drop-off zone offered a sharp rebuke to my bleary-eyed complacency. The airport was packed with mine workers heading up for a two-week shift processing iron ore from the desert mines to the vast shipping port. The flight itself was drowsy and quiet as the majority of my companions returned to their morning slumbers. I spent the flight transfixed, gazing from the window as the wings of the aircraft dipped in a graceful arc into the glowing palette of colors of the landscape below. Upon arrival at Port Hedland I discovered that Hertz offers the budding explorer two types of vehicles: hulking SUVs, with a strong mix of armored personnel carrier in their automotive genes, and pickup trucks rigged for mine operations. To appreciate how I felt maneuvering the SUV out of the airport car park, you should know that at home I drive a twenty-year-old Suzuki Aerio that can outturn a Mini and is best described as "plucky." In my present vehicle by comparison I now felt like a tank commander. Now to run through my desert-driving checklist. Plenty of water—check. Personal locator beacon—check. GPS—check. "In two hundred kilometers, turn right . . ."

Months earlier, when planning this trip from my home base of Victoria, BC, I had made a crucial discovery. Someone even more passionate about stromatolites than I am had written a guidebook. Martin van Kranendonk and Jean Johnston's *Discovery Trails to Early Earth: A Traveller's Guide to the East Pilbara of Western Australia* had everything I needed for a driving tour through the Pilbara: the science, the maps, the photos, and GPS coordinates good to within two meters. Each stop on my driving route to the town of Marble Bar would be marked with a well-researched waypoint.

I had set my GPS alarm to go off within two hundred meters of each point of interest, be it a rock feature close to the road or a vista over the ancient geological landscape. However, I failed to realize that, when driving at 120 kilometers per hour, I was covering two hundred meters every six seconds, and every alarm demanded a rapid slamming on the brakes. Each highlight en route—be it dolerite dike, inverted bed, granitic inclusion, or pyroclastic rock—was testament to the Pilbara's ancient geological past. I was quickly becoming a minor expert, or major bore, regarding the geological history of this stark and austere region of Australia.

If you didn't know that Marble Bar is Australia's hottest town, the giant roadside sculpture at the edge of the town announcing the fact would soon put you straight. The Ironclad Hotel that forms the center of town life is rightly celebrated as an oasis in the desert. When I arrived the barman was bottle-feeding two adopted baby joeys, and my Victoria Bitter—beer to the uninitiated—was served in the can with a cooler sleeve. The real "Bar" in Marble Bar is a great vein of jasper that cuts across the Coongan River on the edge of town. Up close, it is a tilted layer cake of reds, pinks, blues, and grays, each layer having been deposited by hot, mineral-rich water that penetrated and then cooled in the existing limestone rock. The sunset colors only added to the sense of richness and warmth. The air itself was hot—still 35°C at the end of a late winter day. Had I traveled to this part of Australia in the summer, the temperature would be in the high 40s, and I would be in a body bag.

The following day my journey continued to one of the sites that marks the dawn of life on Earth. Sixty kilometers south of Marble Bar my GPS chirped into life, and I hit the brakes once again. My guidebook told me to look for a small rock cairn marking a lightly overgrown four-wheel-drive track—yet before me stretched an unbroken view of spinifex and small trees. (For novices like myself it is worth pointing out that spinifex is a species of grass richly endowed with sharp, silica-infused spines. My long relationship with this prickly plant was only just beginning.) I checked the guidebook again—it was published in 2009—and I realized that I might have to recalibrate for a certain amount of growth over the intervening years. With less naive eyes, I did in the end discern a faint pair of linear features in the spinifex that might be termed a track. It was the best lead I had from a list of one, and I timidly aimed my car off the road.

Fortune does indeed sometimes favor the mildly brave—though the only thing I was being brave with was the paint work on my rental car as I gently "eased" several small trees and shrubs out of the way. A faint jeep trail appeared, followed by the promised gully crossing. Things were definitely looking up, and I celebrated by locating second gear. The trail finally petered out at a small patch of level ground that offered at least a faint hope of turning the car through 180 degrees. I had reached the much vaunted "car park."

On leaving the car I could finally appreciate my newfound isolation. The morning was hot and still beneath an unbroken blue sky. A close silence lay draped across the land, and not a breath of wind disturbed the spinifex dotting the low hills and crags around me. As I climbed the squat hill in front of me and approached the first in a series of innocuous-looking shattered rock outcrops, I felt within me a sensation of walking through time, each footstep transporting me across untold millions of years. Atop a tilted band of limestone exposed in the first of these outcroppings sat a fractured layer of sandstone and chert. This was the ancient reef that I had come so far to see.

These rocks are 3.35 billion years old. How can one even attempt to place a number that large into context? One number from prehistory that most readers remember is that the dinosaurs became extinct approximately 65 million years ago. It is a number fixed in our imaginations along with images of giant, terrible lizards, swampy forests, and flaming asteroids falling from the sky. Yet 3.35 billion years represents an interval that is more than fifty times further back into the past than the time of the dinosaurs, and our minds have no easy mental pictures to fall back on. We have jumped feet-first into deep time—the vast prehistory of Earth where the seemingly arcane language of geologists serves as our only guide.

The age of the rock formation that stood before me is known with tight precision from radioactive dating of volcanic rock layers that sandwich the reef from above and below—squeezing it into a narrow slice of geological time. Heaving waves of force have tilted the whole reef on its side, to be revealed again in cross section at the surface by the inexorable grind of erosion. The rock here is composed of the accumulated sediments that gradually built up over the limestone. Look more closely, and you will see delicate layers within the rock, each perhaps some few millimeters thick. It is these concentric waves and ripples within the rocks that mark out something different and truly remarkable—faint yet discernible traces of living structures.

Nature has many ways of laying down rocks in intricate, entwined layers, from the preserved sedimentary ripples on a sandy riverbed laid down in some eon past to the swirling hydrothermal patterns in the jasper of Marble Bar itself. One should therefore be careful before concluding that some seemingly intricate pattern in a sample of rock points to early life. At this particular location, however, erosion has given us a top-down view of these structures—which are revealed to be squat, conical mounds like so many upturned egg cartons. It is within these layers of ancient sediment that modern microscopy has revealed cell-like structures of carbon and nitrogen atoms—ghostly evidence of single-celled bacteria that made up the mound-like colonies.

To stand before these ancient rock formations and the microfossils they contain is to perform an extraordinary act of communion with our distant past. These are our ancestors—the roots of our family tree, barely glimpsed in the depths of time yet linked to us by an unbroken cycle of cell division and evolution. As I looked at the fossils in front of me, I could feel the years slipping away, all 3.35 billion of them, as my mind's eye took in a view of a forgotten world at once alien and familiar. The scrubby desert in front of me was replaced by an ancient shoreline bordered by a limestone reef, the sharp spines of the spinifex replaced by the lapping waves of a shallow sea. Yet the skies that arched over this scene would likely have appeared a smoggy tone of orange due to the abundance of carbon dioxide and methane in an atmosphere largely devoid of oxygen. The seas themselves, rather than reflecting the blue tones of our modern oceans, would likely have had a greenish tint from the abundance of dissolved iron that floated freely beneath the waves. My imagination had conjured before me this eerie and deeply primordial view. For all I knew, I could have been standing on an alien planet. As I took a step forward, my GPS detected a final waypoint and chirruped into life. "You have reached your destination." Quite. A fleeting moment had passed, and the present world subtly shifted back into focus before my eyes.

It speaks of Australia's great natural diversity that my imagination was aided by an earlier visit to see modern, living stromatolites on the shores Australia's northwest coast. Here, laid out before my actual rather than my mind's eye, stood serried ranks of low gray pillars, formed by successive generations of colonies of primitive bacteria, that stretched out into the calm waters of a vast salty lagoon. These modern-day stromatolites provide a unique living link with the stromatolite fossils of our distant past. Each is a biological time capsule, an untouched relic from the dawn of life that we can walk up to and touch. To return to our dinosaur analogy again, present-day stromatolites help us put living flesh on ancient bones.

As I approached my own encounter with the stromatolites of Hamelin Pool at the southern end of Shark Bay in Western Australia, I felt a growing sense of trepidation and tried not to quicken my pace. In the slanting light of early morning the path wound through low banks of grassy sand dunes to a wide, deserted beach. As the sandy beach gave way to the shore, a close-packed crowd of what appeared to be thousands of gray stone stools clustered on the tide line, apparently advancing as a dark, unified mass into the sunlit lagoon beyond. To one familiar with the evolutionary story told to us by stromatolites they appeared to march off not into the salt-rich waters of the bay but instead into the distant horizon of time itself.

These modern stromatolites are cooperative ensembles of single-celled organisms, with multiple different species sandwiched into millimeter-thin layers. Each squat gray pillar represents a raised island of existence, with the topmost layer typically populated by a surface mat consisting of microscopic strands of photosynthetic bacteria. Oozing a light film of mucus, each of these filamentary microbes slowly glides over its neighbors, intermingling to produce an organically woven mat. These are cyanobacteria, so named for their dark blue color that forms an easily recognizable band when a stromatolite is sliced in cross section. These bacteria are referred to as primary producers—they use the energy of sunlight to combine molecules of carbon dioxide and water to produce simple sugars and oxygen. Their layer in the stratum of life is rich in oxygen, and somewhere deep within the vaults of Earth's history the ancestors of organisms such as these were responsible, one molecule at a time, for suffusing our own atmosphere with the oxygen that would forever change the chemical character of our planet.

Below these penthouse dwellers comes a second layer of either green or purple sulfur photosynthetic bacteria. These are interesting for two reasons. First, their view of the sun is effectively unobscured by the cyanobacteria above them as their particular molecular form of chlorophyll is sensitive to light shifted to the red end of the spectrum rather than the chlorophyll-a used by most cyanobacteria. Thin bands of green and purple reveal the extent of these layers. Second, these bacteria produce organic compounds via anaerobic photosynthesis—without the production of molecular oxygen—and their metabolism offers subtle hints of what kinds of life existed before the evolution of oxygen-producing microbes.

Finally we encounter the basement dwellers— omnivorous microbes that break down the organic sludge that accumulates in this oxygen-starved sublayer. These organisms act to thicken the slimy mucus that binds the whole assemblage together, for within this layer, populations of sulfur-reducing microbes steadily precipitate dissolved calcium carbonate in the form of the mineral aragonite and contribute a further layer of structural integrity to this multistory ecosystem.

The only constants required by living stromatolites are waters that are rich in dissolved calcium carbonate and poor in nutrients that would otherwise promote the growth of algae and smother the stromatolites. A modern stromatolite is therefore an example of well-tuned cooperative living where multiple species of microbes have constructed for themselves tiny but important niches in which to survive. Their importance for our story derives from the fact that the outward structure of stromatolites, whether living or fossilized, has remained relatively unchanged over the passage of eons. This may well be because the niches they occupy, and their biochemical solutions to the challenges of living in them, have likewise remained constant over that time interval. In evolutionary terms they are nature's ultimate players of the long game.

Stromatolites, both ancient and modern, provide us with a glimpse of some of the first steps taken by life on Earth. Yet now our astrobiological story turns to Mars, and we can consider in a new light the growing realization, born from decades of exploration, that the early histories of the planets Mars and Earth show remarkable similarities to one another. The question that then arises is, If life arose on Earth somewhere close to 3.5 billion years ago, could it also have arisen independently on Mars?

Our modern views of Mars and Earth could not be more

different, with a seemingly dead, desert world on one hand and a precious planetary paradise on the other. However, if we retrace the geological history of each planet back through the passage of eons, we obtain tantalizing glimpses of two strikingly similar ancient worlds. Mars and Earth are distant twins—distant in time in this case—and though they were not quite separated at birth, the history of each planet reveals that, from relatively similar beginnings, the slowly grinding cogs that form the gears of planetary evolution eventually drew the two planets down separate paths to form the modern worlds we observe today.

Our window into the early history of Mars is provided by its surface rocks, and that surface is very old. Approximately one-half of the surface of Mars is composed of rocks formed more than 3 billion years ago. In contrast to this, the surface of Earth is considerably younger, with some two-thirds of Earth's surface composed of oceanic crust created at mid-ocean ridges less than 100 million years ago—just one-thirtieth the average age of the Martian surface. In a further difference to Mars, the early history of planet Earth is tucked away in a select few *cratons*—relics of ancient continents that can be found in regions such as Western Australia, where this chapter began. In this sense studying the surface of Mars is akin to discovering a history book lying open on a library desk. There is a great deal written on the open page, and we have read voraciously. But we are fundamentally limited in that we cannot—or at least not without great ingenuity—turn any of the pages to learn about the interior rocks of Mars. In contrast to this, on Earth the great diversity in the ages of surface rocks means we can read many more pages of the book, though only a few of the earliest pages survive intact. Yet whether revealed by high-resolution imaging satellites in Mars orbit or painstakingly scratched from the surface by roving robots, we have discovered that early Mars was wet: 4 billion years ago liquid water flowed in profusion on Mars. From the tentative evidence of ancient shorelines, Mars may even have possessed oceans with depths of several hundreds of meters, and by sorting through the accumulated evidence for ancient liquid water preserved in Martian surface rocks, we can reconstruct a story of an epoch so potentially Earth-like that we name it the Noachian after the great mythical flood that inundated biblical Earth.

The existence of a water-rich early history of Mars has further implications for whether it was a suitable habitat for life. If a large body of water flowed on the surface of Mars today, it would quickly succumb to the twin effects of evaporation and freezing, exposed as it would be to Mars's thin, cold atmosphere. However, the fact that sizable reservoirs of liquid water existed for long enough periods of time in Mars's past to leave their imprint upon the geological record tells us that Mars must have possessed an ancient atmosphere that was much warmer and thicker than that of today.

But what happened to that atmosphere and the surface oceans that it nurtured? The primary evidence for the presence of liquid water on the surface of early Mars is diverse and compelling. However, the corresponding evidence for what created and sustained the atmosphere that allowed that liquid water to persist is much more sparse. We can speculate through an analogy with Earth that a combination of planetary volcanism and a global magnetic field created the conditions for a dense and warm Martian atmosphere. As Mars cooled, as it inevitably would have, we assume that its volcanic activity declined and the magnetic field disappeared. With the gradual loss of its atmosphere that followed these events, we presume that the oceans would have largely evaporated and been lost to space, with the remaining water being locked away as ice as Mars gradually became the frozen desert world that it is today. We risk drifting on a tide of speculation here, cast loose from the secure tether of primary evidence. However, the abundant evidence of ancient oceans, lakes, and rivers on Mars trumps any lingering uncertainty over whether early Mars was habitable. Can we speculate, then, not just that life emerged on Mars over a time scale similar to that on Earth but that it evolved into similar forms? We must clearly exercise caution in bridging the gap between Earth and Mars using the logic of analogy. Yet if we tread carefully, might not the fossil evidence of early life that we have uncovered here on Earth serve as a guide in our search for life on Mars? By taking this step not only can we potentially add a new and valuable perspective to our exploration of early life on Earth, but we can additionally frame the investigation of Mars in the form of a novel planetary experiment: If you start with two worlds, similar in their geological activity, atmospheric conditions, and availability of liquid water, would you expect life to emerge on both planets? If you were feeling especially brave you could speculate further on whether that life would exhibit similar forms on each world.

In contrast to what you might think, such speculation plays an important role in science—but only if it encourages us to try something new, to search for new clues in previously overlooked places. As long as our speculation leads us to open new doors rather than close existing ones, we can consider more deeply how the present evidence of the earliest life on Earth might teach us what kind of evidence to search for on Mars. Yet, in order to unlock the secrets held within Earth's oldest fossils, we will have to zoom in further, much further—down to the scale of individual cells preserved through unimaginable eons. For our story now enters a new phase where the laboratory microscope becomes as critical a tool for the paleontologist as the prospecting hammer was for previous generations. And in the same way that a telescope peering through the cosmos can be likened to a time machine through which we can view galactic worlds long passed, advanced microscopy would soon reveal images of similarly ancient scenes, in this case the Archean world inhabited by early life on Earth.

Geology has favored Western Australia with some of the oldest exposed crustal rocks on Earth. In a very literal sense, it is the land that time forgot. The Pilbara Craton covers a large swath of northwest Australia and represents the exposed bones of an ancient continent whose rocks were originally formed between 3.6 and 2.7 billion years ago. The Archean eon that witnessed the formation of the Pilbara Craton is defined as the period of Earth's history stretching from 4 billion to 2.5 billion years ago. It is bookended in geological terms by the older Hadean eon, during which Earth's violent formation is presumed to have precluded the existence of life, and the younger Proterozoic eon of basic cellular life and the biological production of atmospheric oxygen. Its very name, the Archean, speaks of the "ancient ones." It is the eon that gave rise to the first life on Earth, even if our scientific evidence for that life is sparse and tenuous.

Like petrified bones protruding from the sands of a windblown desert, these exposed Archean rocks contain fossils of extreme antiquity. These are the rocks that contained the preserved stromatolite reefs that I was fortunate enough to visit at the start of this chapter, a particular geologic formation known as Strelley Pool Chert. Given such wonders, it is no surprise that Western Australia has a rich history of geological exploration, motivated by the search for iron ore and other commercial minerals. Yet that exploration has been accompanied by a growing realization that those same rocks provide a unique window into the early history of life on Earth. The path of discovery continues today at the extreme limits of modern microscopy, where one can obtain tantalizing glimpses of the preserved cells that make up these ancient stromatolites. However, we would be wise to recall that earlier in this chapter I counseled you to be wary of the many ways in which nature can lay down sedimentary rocks in ways that mimic the appearance of stromatolites. (In meteorite studies one hears of meteo*rites* and meteo*wrongs*. Sadly, there appears to be no equivalent phrase in stromatolite studies.) One therefore must be careful to look beyond mere surface appearance—morphology in scientific parlance—to search for clues on the microscopic scale.

The very first examples of what was once called Precambrian life were identified in rock samples taken from the wonderfully named Gunflint cherts that form part of the shores of Lake Superior. These smooth black rocks are 2.1 billion years old, and when samples were sliced into thin, transparent sections to be photographed at high magnification, they revealed microscopic, cell-like features held as tiny fossil imprints in the flinty stones. These dark shadows on the photographic negatives, sometimes mere hints of wisps, spheres, and strand-like features, were tantalizing in their resemblance to modern species of microscopic algae and bacteria. Yet at the same time they were profoundly unsettling, given that they were preserved in rocks over 2 billion years old. The discoverers of these fossil traces—Stanley Tyler and Elso Barghoorn—published their results in a 1954 issue of *Science*, then, as now, the journal of the American Association for the Advancement of Science. Their paper, "Occurrence of Structurally Preserved Plants in Pre-Cambrian Rocks of the Canadian Shield," marked the first modern claim for the existence of fossils that predated the Cambrian explosion—the point in the geologic record just over half a billion years ago where fossil life appeared to emerge quite literally from nowhere. Imaging techniques have certainly improved since that first study, and new, remarkable images of ancient life have been discovered. However, the conclusions drawn from that first paper have remained solidly robust over the succeeding years: When sliced thinly and photographed, these rocks have revealed the ghostly traces of clusters of cells. As techniques improved, fine cellular filaments were resolved that, but for their presence in ancient rocks, would have convinced researchers that they were looking at some new species of present-day photosynthetic bacteria. Different, to be sure, yet closely related to modern cyanobacteria. Could careful study of even more ancient rocks tease out a thread of fossil evidence continuing earlier in the history of life?

That would turn out to be easier said than done. Exposed ancient rock formations are rare—there are simply fewer old rocks available to discover. In addition, these rocks have been exposed to a greater extent of geological history. They have been covered by younger rocks, baked at depth in Earth's crust, and ultimately revealed by the incessant grind of erosion. Geological parlance refers to these events as "secondary processing," a term that describes how heat and pressure erase the precious and tiny fossil traces we seek. The challenge is therefore to discover ancient sedimentary rocks that have preserved the microscopic evidence of ancient life yet have been bypassed by the forces of later geological events.

At this point in our story attention began to turn to some of the remarkable geological sites of Western Australia. The Pilbara Craton lies in the northwest of this region of Australia and represents the exposed fragments of a primordial continent. The rock outcrops, riverbeds, and exposed cliff faces of this region provide an unparalleled view of the Archean Earth. Individual locales within the modern Pilbara Shire, with names such as Strelley Pool, Apex, and North Pole Dome, would soon mark the sites of fierce intellectual combat and academic rivalry in the search for the dawn of life on Earth.

The first salvos in what would become an intense scientific debate originated in the Apex chert formation, a band of exposed rock formed just under 3.5 billion years ago that lies about an hour's hike to the west of Marble Bar. William Schopf, a highly respected US paleobiologist, had originally presented evidence of cyanobacteria-like fossils in these rocks as part of a paper titled "Early Archean (3.3-Billion to 3.5-Billion-Year-Old) Microfossils from Warrawoona Group, Australia," published in 1987 with his then graduate student Bonnie Packer. The key evidence presented in the paper were fossil images obtained using optical microscopes and photographic plates—the same techniques Tyler and Barghoorn had used almost thirty years earlier. However, the age of fossils that he and his coauthor had discovered implied a billion-year jump in the history of life on Earth, perhaps to the very first microbial forebears of a long lineage of bacteria that continues to the present day.

Over the next fifteen years Schopf and his coworkers published a number of papers that used new imaging techniques to dig deeper into the mystery of these fossils. Each of their techniques appeared to reveal in greater detail the cellular order of these ancient microbes. Their claim of early life was based upon the premise "I am ordered, therefore I am," a claim spelled out in the pattern of preserved cells traced in these Archean rocks. Many of these articles were published in the prestigious science journal *Nature*, and in reading them one might conclude that the evidence in support of these fossils was compelling and that the case was closed. Yet were you to flip the glossy journal page to the very next article, you would have discovered a paper by Martin Brasier, a UK paleobiologist of equally imposing reputation, titled "Questioning the Evidence for Earth's Oldest Fossils," which cast doubt on the claims of Schopf and his team and suggested instead that the purported fossils were merely microscopic artifacts of carbon deposited on individual mineral grains in an ancient hot spring environment.

Although their now notorious academic feud would continue over the years, it masked a deeper clash of scientific philosophies that, as astrobiologists, we would do well to note. Schopf essentially stated that, given our best set of criteria for what early life should look like, the Apex fossils passed this threshold and should be accepted as evidence for early life. Brasier instead took the view that early life should be the explanation of last resort and that one should exhaust all alternative, nonbiological explanations for one's evidence before making such a claim. These differing approaches to proof might seem (quite literally) academic; however, these ideas also form the basis of two pillars of many legal systems around the world. In a criminal court, for example, the prosecution must convince the jury of the defendant's guilt beyond any *reasonable* doubt—a high burden of proof required by the threat of criminal penalties. On the other hand, in a civil court the plaintiff must simply present enough evidence to convince the judge or jury that there is a greater than 50 percent chance (for example) that they are right—a so-called *preponderance of evidence*. Returning to the question of early life, we can therefore observe that Schopf had indeed presented a preponderance of evidence. However, Brasier had also raised reasonable doubts. It would be satisfying to state that one side was right and the other wrong, but that is rarely the way science works—especially at the cutting edge, where there may be insufficient data to reach a firm conclusion.

The debate between the two teams would rage for many years, with each camp drawing on new evidence and new laboratory techniques to either advance their own ideas or form a new defensive line against doubters. Blow and scientific counterblow were struck. Brasier and his scientific collaborators would go on to stake their own claim in the early life debate by publishing a paper arguing that a nearby locale of comparable age to Apex contained more convincing evidence of microfossils and ancient metabolic activity. Their paper, "Microfossils of Sulphur-Metabolizing Cells in 3.4-Billion-Year-Old Rocks of Western Australia," used a new instrumental technique to literally blast atoms off the surface of each fossil sample and map the ejected ions as they scanned across each slice of rock. Their images revealed micron-sized spheres rich in carbon and nitrogen atoms. Could these be the ancient remains of individual cells? Moreover, in a tantalizing glimpse of the possible life cycle of these Archean cells, many were found to be clinging to the edges of tiny mineral grains rich in sulfur. Since the processing of sulfur is suspected to be a key metabolic

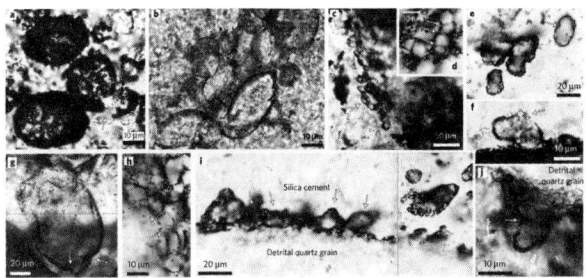

Ancient terrestrial microbes Are these fossilized microbes in 3.4-billion-year-old rocks or microscopic wisps of a fertile human imagination? These electron microscope images show thin slices of ancient reef rocks known as the Strelley Pool Formation. Scientists are still grappling with the question—both on Earth and on Mars—of how to distinguish the signatures of ancient microscopic life from nonbiological deposition of mineral crystals. Credit: David Wacey et al., "Microfossils of Sulphur-Metabolizing Cells in 3.4-Billion-Year-Old Rocks of Western Australia," *Nature Geoscience* 4 (2011): 698-702.

pathway that emerged early in the history of life, these images had potentially revealed a primordial cellular tryst.

Sadly, Martin Brasier passed away in 2014, and the scientific community lost an original and highly capable thinker. The debate concerning the veracity of Earth's oldest fossils continues in the journals of today, yet some of the academic intensity that accompanies such strong disagreements has been lost. In one respect the Schopf-Brasier debate resulted from the MOFAOTYOF principle. As Brasier explained in his excellent book *Darwin's Lost Worlds*, this stands for "My Oldest Fossils Are Older Than Your Oldest Fossils" (equivalent principles exist in nearly all scientific disciplines). In another respect, though, their debate was important because it was *productive*. Each side sought new data and developed new laboratory techniques to support their respective claims. The jury may still be out regarding many of the individual fossil claims put forward by each side, but that is, I think, more of a reflection of the truly challenging, and therefore truly interesting, occupation of interpreting fossils of the greatest antiquity.

The debate over Earth's oldest fossils is clearly reminiscent of the earlier debate over claims that the Martian meteorite ALH84001 contained evidence of life on the red planet, and both episodes should stand as sobering reminders to those who think that returning ancient Martian rocks to Earth represents the sole challenge involved in answering the question of whether life arose early in the history of Mars. We have learned on Earth that claims of biogenicity—proof of life—must be supported by convergent lines of evidence that can be weighed and balanced against nonbiological explanations according to rules that are accepted by a quorum of scientists in the field. If we do not yet possess either the data or the techniques to answer this question definitively when applied to some of Earth's oldest rocks, we can only approach the prospect of attempting to do so with Martian rocks of similar age with a sense of trepidation. Excited trepidation to be sure, but nonetheless we must remain aware of the tremendous uncertainties that await us. For it is to Mars that we must now travel, crossing an orbital gulf of almost 75 million kilometers to sample rocks of ages similar to those of the Archean Earth and drawn from environments astonishingly reminiscent of the ancient marine worlds that we have so far explored on Earth.

Mars has been visited by five NASA science rovers since 1997. Their names and achievements make for an impressive roll of honor: *Sojourner* (1997), *Spirit* and *Opportunity* (2004), *Curiosity* (2012), and *Perseverance* (2020). Each of these missions has been constructed upon the engineering foundations of its forebears, and each has blazed a trail of scientific discoveries for its successors to follow. In combination with orbiting observatories and static landers, these rovers have provided twenty-eight years and counting of continuous scientific activity at Mars—an astonishing achievement. As a result of their endeavors we know the present conditions and ancient history of Mars in greater detail than ever before. Yet the fundamental questions persist: Did life ever exist on Mars? Does it exist today?

The most recent of these rovers, *Perseverance*, was the first to be given the specific task of helping to prepare for a future mission to return samples of Mars rocks to Earth. For many scientists that goal represents the next logical step in Mars exploration, as a robotic sample-return mission is much cheaper and involves far less risk than human space flight. Sample return also offers a critical capability that all one-way missions lack—flexibility. The science performed by all spacecraft, wherever they boldly go, is limited to their suite of onboard instruments. Each probe, whether orbiter, lander, or rover, can be thought of as a compact and lightweight laboratory where every piece of technology represents a superbly miniaturized clone of a common scientific instrument. However, reducing the size and weight of the science package often comes at the cost of reducing the sensitivity and features of those instruments. Moreover, the scientific wish list of ideas jostling to board the flight to Mars is always more than can be accommodated on the tightly packed spacecraft. Each winning idea is selected according to a strict set of prelaunch science goals, but that still means many good science ideas are left behind. If we then fast-forward to our days of exploration on Mars only to discover a rock that requires a test using just the very instrument that failed to make the prelaunch cut, or if we excitedly note a trembling signal that requires an instrument of fractionally greater sensitivity, then our moment of discovery would crash against the hard reality of robotic space exploration. By returning rocks to Earth, a Mars sample-return mission offers scientists the critical opportunity to pursue a tantalizing or marginal result with an unrestricted range of laboratory techniques, bypassing the limitations imposed by the fixed package of science instruments carried by one-way missions.

A sample-return mission to Mars does not come cheap, however, since it embodies multiple space missions in one. Not only must you dispatch a spacecraft to the Martian surface, you have to bring part of it back safely to Earth. The additional steps required to return samples to Earth bring with them significant new engineering challenges, perhaps foremost among them the milestone of conducting the first rocket launch from the surface of another planet. Not surprisingly, the broader scientific community has justifiably questioned the high cost of such a mission. It is a perennial debate within science: Should we concentrate research funds into a single ambitious mission at the expense of perhaps sending a second mission to an equally compelling planet or moon elsewhere in the solar system? Yet if *Perseverance* were to drill a precious few grams of Martian rocks and lay them up in a cache for a future mission to return to Earth, those rocks would be scrutinized more precisely than any before them. And if we have learned one thing from studying Earth's oldest fossils, it is that new discoveries always lie at the very limits of what our instruments can achieve. However, if using the present Mars rover *Perseverance* to select rock samples for return to Earth is the first stage of a future sample-return mission, the question then becomes, What rocks will be given the golden ticket for the trip back to Earth, and where must we send the rover to find them?

It will come as no surprise to learn that NASA has a thorough process to determine where on Mars each new generation of rovers will be dispatched to explore. First there are the basic engineering requirements: Where is it safe to land, and where is it efficient to operate? During a rover's plunge to the Martian surface, it deploys a parachute to slow and control its descent. Therefore, wind patterns in that region of the planet must be sufficiently well-behaved to paint an imaginary bull's-eye on the surface that corresponds to the rover's most likely landing area. Open, smooth terrain provides the best prospect for a trouble-free touchdown, and the area within the landing ellipse will have been scoured for any obstacles larger than twenty-five centimeters across by the electronic cameras of the Mars Reconnaissance Orbiter (MRO). Of course, one could always send the new rover to terrain that had already been surveyed from the ground by one of its forebears, but this would mean missing the vital opportunity to explore new territory and, one hopes, make new discoveries. Once on the ground, the rover then has to traverse the Martian terrain toward its science objectives, and once again the MRO will have mapped out the likely routes, with engineers balancing the required travel time with the estimated mission lifetime.

To determine exactly which regions on the surface of Mars would offer the greatest scientific promise for *Perseverance*, NASA held a series of prelaunch site-selection workshops where scientists, both within NASA and from the academic community in general, could pitch and debate their preferred spots. Over a number of such workshops, the potential landing sites were whittled down from thirty-four to eight, then from eight to three, and from three eventually down to one. The overarching science goals to be met by the chosen site were set forth by NASA in a series of questions: Where is there the prospect of discovering rocks that might once have harbored life and that additionally might have preserved evidence of its presence? Where might there also exist rocks that will allow scientists to reconstruct the geological history of such a habitable environment? Is the selected region suitable for a later sample-return mission, or might it be possible to learn clues about Mars that would be of use to later human exploration?

A region within the Jezero crater emerged from this process as the primary landing site for the *Perseverance* rover. *Jezero* is a Slavic word for a lake, and the ancient crater was once filled with water to a depth of up to 250 meters. The route to be taken by *Perseverance* traverses a beautifully preserved river delta some 3.4 to 3.8 billion years old, in addition to the surrounding lakeshore, and the rover will eventually surmount the crater rim itself. Orbiting satellites have detected the signature of clay minerals within the ancient lake bed that raise the possibility that Martian fossils might have been preserved in their fine-grained texture. Furthermore, remnants of volcanic deposits offer a chance to determine accurate radiometric dates for the whole assemblage. Given this ancient, water-rich landscape, even if we were to put aside the possibility of a sample-return mission, the prospects for scientific discovery at Jezero are truly exciting. *Perseverance* landed at the Jezero crater on Mars on February 18, 2021, and began a new chapter in the history of Mars exploration. From our perspective, driving a rover across the Martian surface offers the budding astrobiologist an unmissable opportunity to explore

Mars from the comfort of an Earth-based lab. What does a day in the life of *Perseverance* and its human controllers look like to a visitor to the Jet Propulsion Lab in Pasadena where the science team is based? The sign that welcomes you to JPL tells you to "Dare Mighty Things," and this lab is indeed one of our planet's most important launchpads, where human ideas take flight into the solar system and beyond. For the *Perseverance* team within the building, the daily rhythms of life are set by two overlapping cadences. One is the daily cycle of uplink, downlink, and science planning. These activities form the core of the *Perseverance* mission, and the three teams occupy adjoining control rooms at JPL. The other is the twenty-four-hour and thirty-seven-minute cadence of the Martian day, or sol, which slowly creeps into and out of sync with the Earth day over months of operations.

The uplink team is in charge of sending instructions to the rover, and these commands form the nuts and bolts of its operations. Each wheel rotated, each camera image captured or science sample analyzed, occurs in response to commands sent to the rover, line-by-line, in a careful sequence of computer code. If the uplink team can be thought of as being responsible for talking to *Perseverance*, then the downlink team is tasked with listening to it—to every byte of telemetry it transmits, be it the torque on the wheels, the images of the landscape it captures, or the mineral composition of a rock sample its instruments obtain. At the heart of all of these operations is *Perseverance*'s onboard computer, and although in terms of speed its raw specifications would not make even a mid-nineties desktop blush, it can robustly operate during the frigid Martian nights, and its memory chip can tolerate the damage caused by unhealthy doses of cosmic radiation. So vital is this computer to *Perseverance*'s operations that the rover even carries an entirely redundant spare—just in case.

An unavoidable challenge of Mars exploration is the time it takes to communicate with the remotely operated rover. The ROV pilots I sat next to on the *Nautilus* were connected to *Hercules* and *Argus* by a nearly instantaneous link to the seafloor, and the time it took for each command to travel through the optical fiber to the ROVs was imperceptible to us humans. Although the radio waves used to communicate between the science team at JPL and *Perseverance* on the Martian surface travel at the speed of light, the distances over which the signal has to travel and then return are much, much greater. Even when perfectly aligned, it takes a radio signal a total of eight minutes to travel from Earth to Mars and back again. However, the two planets' relative positions rarely line up to produce a time delay that short, and the round-trip time delay can be anywhere from eight to forty minutes. Yet even this interval is typically much shorter than the actual communication lag with Mars.

In practice, uplink and downlink events occur just a few times each day. The backbone of NASA's solar system communications is the Deep Space Network (DSN), an array of three giant radio antennae separated by approximately 120 degrees of longitude on the globe that provide a constant view of spacecraft as they explore the solar system. Although the rover can communicate directly with Earth using the DSN, it is a shared network, operating at the relatively low data rate of up to thirty-two kilobits per second. Most readers will know the arguments that ensue when their family attempts to share an overloaded home Wi-Fi network; in comparison to this, the task of trying to coordinate the various space missions communicating over the DSN is similar to a large family attempting to share the data bandwidth of a 1990s-era dial-up modem. All of this assumes that the portion of Mars occupied by the rover is pointing toward Earth and that onboard power is available for the radio signal to reach Earth. A more effective option available to *Perseverance* is to use a Mars-orbiting spacecraft such as MRO as a communications relay. Not only does this require less broadcast power, but data can be transmitted as fast as two megabits per second. The only catch is that *Perseverance* can transmit to an individual orbiting spacecraft for just eight minutes a day, and it has to be ready and waiting with a package of data at such daily broadcast opportunities.

The cumulative effect of these limitations is that the *Perseverance* team at JPL is restricted to a handful of communications windows with the Mars rover each sol. These windows represent the smallest denominations of communications currency available to the scientists and engineers tasked with managing the pace of the mission. Perhaps the easiest way to appreciate the effect of these communication lags is to imagine yourself driving the rover across the surface of Mars. Many of us have played video games, and some have piloted personal drones or remote-controlled vehicles. With the benefit of onboard cameras and instantaneous communications, one can achieve an incredible sense of remotely enabled presence, of flying with the drone or racing around a track. However, to imagine the effect of a communications delay when maneuvering a Mars rover, consider how your own driving style would change if you were only able to open your eyes once every four to five hours. Your progress across even the least challenging terrain would slow to increments of a cautious meter here or a sedate turn there. Your sense of care might well be heightened by the knowledge that your vehicle cost just over $2 billion and doesn't carry any insurance.

In an age where we are starting to see the first self-driving cars, it will come as no surprise that NASA is leading the way with off-world autonomous driving. *Perseverance* has a suite of navigation cameras to help it map out safe routes across the Martian terrain and a dedicated onboard computer for image processing and visual odometry that constantly checks and references the rover's progress. *Perseverance* has even used its free-flying drone, *Ingenuity*, to scout ahead and take images of challenging terrain along its path. With the aid of these automated navigation tools, the rover can occasionally advance by figurative leaps and bounds, with its personal record having been to cover 167 meters of terrain in a single Martian sol.

At this point one can appreciate why remote Mars exploration proceeds at a pace akin to that of a stop-motion animation. On the plus side, however, is the reassurance that you have a lot of time available for exploration—*Perseverance* has been active on Mars for more than 1,700 sols, and its predecessor, *Curiosity*, has been rolling strong for more than thirteen Earth years. *Patience* might have been an appropriate name for any of these rovers, since it is a key virtue the mission team needs, in addition to the scientific discipline required to make sure that the few activities that can be scheduled each day fit seamlessly into a wider science plan that might take months or even years to achieve.

The third team at JPL, science planning, is tasked with making sure that, on any given sol, the rover knows what to do and where to head and—most importantly—that each and every operation contributes to the overall science goals of the mission. It is not an easy task, since not only does exploration proceed one scoop, drill core, or laser blast at a time, but—more importantly—the mission is at its heart one of discovery. Interesting surprises and unexpected revelations are the daily bread of Martian explorers. Yet how should one deal with the urge to investigate that compelling boulder sitting ten meters away in the opposite direction from tomorrow's drive plan? Or the even more frequent situation where scientists are clamoring to make just one more mineral measurement with one of *Perseverance*'s suite of onboard instruments before heading off on the road again?

Some team members are seasoned veterans who appreciate that sacrifices must be made in order to stay on schedule as the rover inches toward a distant yet significant goal. Others may find in every rock that appears in *Perseverance*'s cameras an unmissable scientific feast that has to be devoured. The path to success lies in striking a balance between the differing views of the science team, because if you have assembled your eclectic and passionate members with sufficient care and forethought, then there definitely will be differing views. That's the whole point. It is perhaps not too far from reality to imagine a schoolteacher taking their elementary science class to Yellowstone National Park for the day. Surprise and awe lie around every bend in the path, and one of the gifts of childhood is to find wonder in the marvels that lie beneath a nearby leaf or skulk at the edge of a riverside hot spring. The challenge for the teacher is to keep the day on track and make sure that everyone makes it back to bus when the sun goes down. The main difference, of course, between our class of fourth graders and the *Perseverance* science team is the abundance of graduate degrees in engineering and planetary science. However, the point of similarity is the need to foster curiosity and remain nimble enough to recognize a good clue when it comes along.

Perseverance's main science goal is to explore the Jezero crater for evidence of past habitability. As it does so the rover has been steadily accumulating rock samples for a future Earth return mission. Perseverance carries with it forty-three sample tubes for its journey across Mars, with each tube consisting of an approximately six-inch-long cylinder of titanium that weighs some fifty-seven grams. Upon drilling a finger-sized core sample of rock or regolith, the material is slid into the tube and then sealed and cached within the rover's chassis. Five of the sample tubes are filled with inert material designed to capture atmospheric or particulate matter and bear witness to the background environment local to each sampling site.

As of December 2022, *Perseverance* had filled eighteen sample tubes. Although the rover itself is the principal repository of sampled material, NASA wanted a backup plan in case some accident left *Perseverance* unable to be located by a future sample-return mission. To that end the rover spent just over one month laying a careful cache of the sample tubes that it had filled so far. Rather like a bizarre mechanical beast laying cylindrical eggs, the rover deposited each of the sample tubes on a flat piece of ground some thirty meters across. Each tube was imaged and tagged at a precise position so a future robot explorer could recover them if need be.

With good fortune, patience, and, of course, perseverance, the rover and its team of scientists and engineers will have many years of exploring ahead of them. Yet it would be remiss of us to miss this opportunity, from our remote vantage point located on a rocky outcrop on Mars, to gaze into the future and imagine NASA's next mission to the red planet, which will return these samples to Earth. To many solar system scientists, sample-return missions represent the holy grail of space exploration. To travel to a distant location in the solar system and bring a little piece of it back with you offers the unrivaled opportunity to perform science beyond the confines of miniaturized onboard laboratories. Humans and robots have returned lunar material from our own moon to Earth, and space probes have visited comets and small asteroids to prospect for a few precious dusty grains to return to Earth. But no mission has yet returned samples to Earth from the surface of another planet.

The main reason for this is gravity. The small asteroids that have been the targets of robotic sample-return missions such as NASA's *OSIRIS-REx* and the Japanese space agency's *Hayabusa2* have puny gravitational fields in which a single spacecraft can gently approach the asteroid, scoop, and then depart with the smallest of nudges from onboard thrusters. As we will learn when we tell the story of these missions later in the book, even this relatively simple sequence requires careful choreography and elegant engineering to achieve success. However, the force of gravity is much stronger on Mars than on these kilometer-sized space rocks, and NASA has demonstrated that it takes a combination of a heat-shield, parachute, and rocket crane to get a beefy rover such as *Perseverance* safely onto the surface. Landing on the surface of Mars incurs a significant debt to the force of gravity, and at the liftoff back to orbit that debt must be repaid by a capable and reliable rocket. The challenge of launching an ascent rocket from the surface of Mars back into orbit with a cargo of Martian rocks will demand new levels of daring from the mission scientists and engineers.

Realizing the need for a coherent, long-term approach, in 2018 NASA and its European partner ESA (the European Space Agency) began the formal process of planning a joint sample-return mission to Mars that would exploit the groundwork undertaken by *Perseverance*. NASA would be responsible for landing the three-ton sample-return lander within a sixty-meter bull's-eye on the surface of Mars. The plan would see *Perseverance* parked nearby, ready to drive up to the lander and hand off its samples to a European robot arm. The next step represents the crux of the entire mission—sending the sample container into Mars orbit atop a solid-fuel ascent rocket. This is indeed the point at which the design team will need to dare mighty things, as no rocket has ever launched remotely from the surface of another planet. The relay race would then move into low Mars orbit, where the baton of rock samples would be passed to a European Earth return capsule. No matter how simple it may appear when written down on the page, this series of what are essentially three separate planetary exploration missions—lander, ascent stage, and return capsule—represents a unique and ambitious undertaking. Each step in the chain of custody is linked by the common engineering requirement to safely pass the rock samples along each leg of their journey to Earth.

If successful, a Mars sample-return mission will mark a generational moment in the exploration of our solar system. The last two community-led, decadal reviews of US planetary science have, quite rightly, put a Mars sample-return mission firmly in first place. However, the most challenging part of the mission may well be to secure the necessary funds to allow the spacecraft to leave planet Earth in the first place, let alone bring it back from Mars. In 2023 a review board commissioned by NASA to oversee the project concluded that the mission cost had doubled—from just over $5 billion to an eye-watering $11 billion budget. In response to this, NASA administrator Bill Nelson told the Mars sample-return team that enough was enough—that costs were too high and the time scale was too long. Nelson challenged NASA and the wider science community to come up with cheaper and faster alternatives. From the perspective of the NASA science teams tasked with planning other solar system missions, whose budgets a Mars sample return was threatening to consume, Nelson's decision came as a welcome relief. Projects such as the *Dragonfly* mission to fly a rover on Saturn's moon Titan, or even plans to return to Enceladus or Europa, were in danger of being pushed back later and later as Mars devoured NASA's planetary exploration budget. From another perspective, however, the physical requirements of a Mars sample-return mission remain unchanged—you still have to lift rock samples from the surface to low Mars orbit and reach the European return capsule. With the science payback from a sample-return mission still preeminent, NASA appears committed to the project but finds itself at a fork in the road. To the right lie cost overruns and budget woes. To the left lies uncertainty. Perhaps the success of the *Ingenuity* drone, which far exceeded expectations during its test flights on Mars, will provide a cheaper and lighter method of returning the sample tubes to the ascent rocket. One can only hope that the promise of a generational science achievement gives the team members hope during the long dark night of these funding challenges.

 Delays and budget woes are nothing new in challenging space exploration. NASA's James Webb Space Telescope is rightly lauded as having established a new frontier in astronomical research, yet even it had more than its fair share of setbacks en route to its eventual launch on Christmas Day 2021. Sometimes the destination can be worth the difficult journey, and Mars sample return is a very exciting goal. The *Perseverance* science team has shown dedication, care, and insight in their choice of sample sites. The key question, however, may well turn out to be, Have they been lucky? Will evidence of ancient Martian life be found lurking within the mineral matrix of a sedimentary rock? Might it be hiding in plain sight, betrayed by an anomalous ratio of chemical isotopes? These are the wonderful uncertainties that are part and parcel of the life of an explorer. Am I confident that Martian rocks will be returned to Earth in my lifetime? I will admit that the road to Mars and back is proving to be longer and harder than I and many others believed. Yet for such an ambitious project to succeed, both the scientists that construct the mission and the governments that fund it—backed by taxpayers such as you and I—need to be united in their daring vision.

So there you have the story of Earth's oldest fossils and their ancient link to Mars. From a lonely rock outcrop in Western Australia one can gaze in two separate directions and take in two different views of utter immensity. One direction provides a view through time, to an Archean reef on the shores of a shallow sea where the horizon fades to the hazy slopes of a giant shield volcano. The use of the phrase "deep time" falls far short as a description of the terrible gulfs that one's imagination must attempt to bridge. Instead we fall back on terms such as Precambrian, Proterozoic, and Archean—waypoints on the path to the first life on Earth. The other view takes in a red planet shining in the night sky. The gulf in distance between Earth and Mars is neither greater nor lesser than the gulf of time that separates us from the early Earth. We cross each one with the aid of technology that is both impersonal and detached. The views of a Martian landscape provided by a robotic rover are neither more nor less personal than the electronic image of an ancient terrestrial cell captured by an advanced microscope. One day, I hope sooner rather than later, we will return Martian rocks to Earth and ask of them the same questions that we have asked of some of Earth's oldest rocks. Two ancient and long separated twins will be reunited. And we will learn whether they share their stories of life.

At the end of my own journeys I neither swam with the stromatolites at Shark Bay nor did I collect any of the wonderful fossils I observed on my travels through the Pilbara. There would be little point in telling you about the scientific wonders to be experienced at these locations if everything you discovered upon traveling to them was eroded and damaged. Therefore, should you be fortunate enough to find yourself treading some of the paths that I have followed in this chapter, step carefully, take many photographs, and, in the best interests of any passengers in your vehicle, set your GPS alarm to at least five hundred meters.

4

THE ARC OF THE FIRMAMENT

MAPPING EXOPLANETS

The rich colors of sunset herald the end of a clear winter's day on the high slopes of the Andes mountains. My vantage point on the metal catwalk that circles the outside of a telescope dome at the La Silla Observatory in Chile looks over an unparalleled horizon. As the sun descends to the Pacific Ocean in the west it casts long shadows over the foothills that recede into the distance below me. Blue shades to mauve and then deepest purple in the hollows of the landscape. The sky itself flames with red about the descending sun, fading to a steely blue as one takes in the vault of the sky from west to east.

The astronomers with me on the exterior catwalk of the telescope head inside, eager to commence the symphony of the night. As the curtain of dusk is slowly drawn across the sky, the first stars appear, seeming to wink into existence against the fading fabric of scattered sunlight in our atmosphere. In the evening sky of classical antiquity, the first stars to appear were labeled as being of the first rank or magnitude. As the sky darkened, the stars of second magnitude would appear, followed all the way to those of sixth magnitude, the faintest stars that could be seen with the naked eye on a dark and moonless night. It is an elegant approach to measuring stellar brightness that comes from an age when the human eye was the sole means of detection.

Beyond the open dome of the telescope the sky beckons, paling rapidly from the day yet still half an hour or more from true night. We use this brief window of time to perform a few final calibrations before we begin our journey across the sky. True night occurs at astronomical twilight, when the sun is eighteen degrees below the horizon and the sky has darkened to the point where it will now change little before dawn.

We take our seats in the control room with the practiced coordination of members of an orchestra taking their positions before the evening's music begins. And there are indeed aspects of performance reflected in the rituals practiced throughout the night's observations. As the stars wheel across the sky, our telescope follows them in closely choreographed motion. Mechanical gears mesh with a computer model that traces the arc of the firmament, the paths formed by the stars as the Earth rotates beneath them. Rotation and counterrotation fix the gaze of the telescope on a single steady point of light. The observatory dome is open to the night, and within it the telescope itself, immense in the darkness, stands as a mute witness to the imperceptible passage of the stars. At the precise instant when all is held in alignment, a computer command opens the shutter of an electronic camera. A slow trickle of photons drips electric charge onto a silicon detector and creates a digital snapshot of a distant star, a single pinpoint of light in the universe. A perfect moment captured in space and time.

From our own vantage point on the pale blue dot, the night sky offers humanity both a gift of wonder and a profound challenge. Ever since *Homo* has been *sapiens* we have stood or sat beneath the stars, deep into the primordial night of our ancestry. We have told stories about the stars and we have studied them, each myth or theory constructed in order to understand our place beneath such majesty.

Truly dark skies, which unfortunately we only rarely encounter today, are deeply encrusted with stars—glittering jewels forever suspended beyond our reach. Science has long pondered their nature, and, although distant and aloof from the hubbub of our solar system, we have come to know their life stories in wonderful detail. Humans have long cherished the belief that the stars would nurture planetary systems like our own, yet only in the past thirty years has that dream of exoplanets become reality.

However, that discovery has brought humanity to the cusp of a very profound question: Do these planets themselves play host to life—whether like that of our own Earth or life unlike our own on truly alien worlds? Sometimes the wonder associated with science comes from the simple act of being able to ask a particular question, to express a new idea or possibility within the scope of our comprehension. That is what we are currently doing by asking whether life exists on exoplanets.

And we have come a long way in a short time. The first planet orbiting a normal star like our own was discovered in 1995. One became a few, and a few became tens. With the launch of dedicated space missions such as NASA's *Kepler*

Pale blue dot The "pale blue dot" image taken on February 14, 1990, by the *Voyager 1* spacecraft at a distance of six billion kilometers. Planet Earth is just a few pixels across and appears as the tiny speck of dust highlighted in the shaft of light. Do we represent a single mote of life in an empty universe, or are we instead just one among countless examples of life in the universe? The science of astrobiology aims to find out. Credit: NASA/JPL-Caltech.

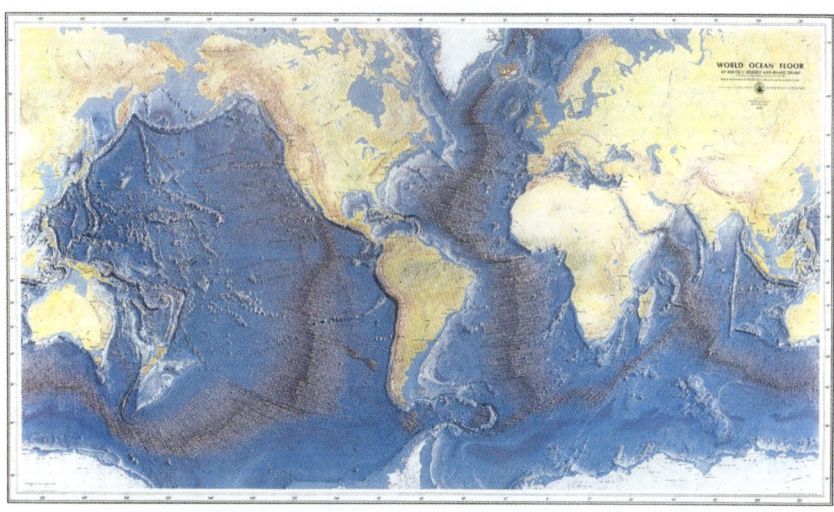

Heezen and Tharp map, 1977 *(above)* A physiographic map of the sea floor produced by Bruce Heezen and Marie Tharp in 1977. Like the stitching on a baseball, the mid-ocean ridge system forms the geological seam of planet Earth. Credit: Marie Tharp Maps LLC. Reproduced by permission of Fiona Schiano-Yacopino.

Hercules launching *(right, top)* Once more unto the deep. The remotely operated vehicle (ROV) *Hercules* is gently lowered from the deck of the E/V *Nautilus* into the waters of the northeast Pacific Ocean. The Endeavour hydrothermal vent systems lie in wait some two kilometers under the ship's keel. Credit: Ocean Exploration Trust.

Endeavour vents *(right, bottom)* A forbidding, alien landscape. A black smoker chimney photographed at Endeavour by ROV *Hercules*. Without the lights from the ROV, the scene would be utterly dark and only some 2°C above freezing. Despite the harsh conditions, the life that abounds in this environment may be the closest analogue on Earth to the creatures that may live inside the icy ocean moons of Europa and Enceladus. Credit: Ocean Exploration Trust.

Stromatolites Seeming to march off into time itself. The living stromatolite reefs of Shark Bay in Western Australia are the modern descendants of a 3.5-billion-year-old lineage of microbial life on Earth. Had the ocean been green and the sky orange, one could have been tricked into thinking one was standing on the Earth of deepest prehistory. Credit: Author.

Ancient stromatolites The eroded remains of sediments trapped by ancient microbial mats that lived 3.35 billion years ago. It is an astonishing act of communion to stand before what is some of the earliest evidence of life on Earth. Credit: Author.

La Silla Observatory *(above)* La Silla Observatory in the foothills of the Andes Mountains in Chile. The dome of the 3.6-meter telescope dominates the foreground, and my old telescope, the New Technology Telescope, looks like an immense aluminum shed just farther down the ridge as we view it. The TRAPPIST telescope, which discovered seven rocky worlds in orbit about a single red dwarf star, is lost in the cluster of small telescopes at the far end of the ridge. Credit: ESO, CC BY 4.0.

The ELT taking shape *(right, top)* The behemoth of the Andes. The structure of the Extremely Large Telescope under construction at Cerro Armazones in northern Chile. The vast dome will house a mirror just under forty meters in diameter. Will it be the first to glimpse the exhalations of alien life in an exoplanet atmosphere? Credit: ESO/G. Vecchia, CC BY 4.0.

Meteorite hunting *(right, bottom)* The stony plains of Morocco provide an open canvas upon which to pick out the dark fusion crust of meteorites—at least that is the idea. However, it was going to take more than one week in the Sahara to discover my own hoard of space rocks. Perhaps my companion in the distance was having better luck? Credit: Author.

Dolphin research Atlantic spotted dolphins are among the most compulsive communicators on planet Earth. The image here captures a lively interaction with bottlenose dolphins in the clear waters of the Bahamas. Can eavesdropping on their vibrant vocalizations teach us language lessons that we might one day apply to an alien message? Credit: Wild Dolphin Project.

observatory in 2009, tens became thousands. The starkest realization from this first age of planetary discovery was one of variety. Every shape or form of exoplanetary system that could exist did. Multiple planets around a single star? Check. Planets in orbit about tightly bound binary stars? You've got it. Rocky worlds orbiting within the "Goldilocks zone" of habitability? Very much so. Hot Jupiters and Super Earths? Nature appears to have it all.

We are now moving beyond this first age of planetary discovery and are beginning to tell their physical stories. What types of planets are most common, and why? How are they born? What forces shape them? Yet one question still dominates all others, a question that seems to motivate the entire field of exoplanet studies: Do they host life?

The idea of the pale blue data point changes as we move from the previous chapters to this one. We are no longer seeking analogues of alien worlds here on Earth. Instead, we are performing astrobiology while keeping our feet squarely planted on terra firma. In this sense planet Earth is our front-row seat to the rest of the universe—though, in saying that we are in the front row, we are certainly overstating things. We are actually sat pretty far back from the stage, and we will need some powerful telescopes to bring those distant views into sharper focus.

I have been fortunate to have been a professional astronomer throughout this age of discovery. I started graduate school at the Institute of Astronomy in Cambridge in 1995, and in 2000 I began my first real job as a researcher and observatory scientist in Chile. During my three years there I spent over two hundred nights beneath the stars, observing with all of the major telescopes located along the Andes. In this chapter I want to tell the story of what it is like to use these incredible telescopes to answer some profound questions about our place in nature. How does it feel to experience the rhythms of life at an observatory? Is it all profound majesty, or do professional astronomers suffer their fair share of failed observations and frustrating nights? I began this chapter with my memories of just one night among the many I spent observing the night sky. I served my apprenticeship in Chile at the New Technology Telescope (NTT) at La Silla astronomical observatory. The observatory is set in the dry foothills of the Andes where, at 2,400 meters above sea level, a long, saddle-like ridge provides a home to a string of white and silver domes. A two-lane access road winds between the telescopes as it traverses a friable landscape of rocky crags and scrubby vegetation. Each telescope that we pass on the way has a story to tell in our search for exoplanets. The small dome of the old Swiss T70 telescope is now home to a Belgian-led survey that discovered a diminutive solar system of seven Earth-sized worlds in orbit about a seemingly unremarkable dim red star. The whole ensemble turns in orbit within a region smaller than the orbit of Mercury in our own solar system. At the summit of the ridge stands the great white dome of the old 3.6-meter telescope, now home to a high-precision spectroscopic instrument that has discovered some 130 new worlds in orbit about distant suns—the largest number discovered by any ground-based telescope.

The turnoff for the New Technology Telescope appears just before the road climbs the last crest before the summit. Though now a venerable telescope—it saw first light in 1989—the NTT served as a test bed for a number of technological innovations that paved the way for later generations of giant telescopes located at the northern end of Chile's Atacama desert. And although our journey in this chapter will take us to those newer and larger observatories in northern Chile, here is where my own story began. Telescopes such as the NTT are the reason why the control room of the *Nautilus* felt strangely like home—the cramped human spaces in front of banks of monitors, the hum and vibration of a mass of technology, even the gentle sway as the entire dome structure would rotate to point the telescope to a new location on the sky.

At the very start of our story of astrobiology, we encountered the prescient words of Giordano Bruno. Writing in 1584, he speculated that, even if the stars of the night sky were accompanied by planets of their own, they would remain hidden from view by the immense glare of their parent stars. The story of how we came to discover exoplanets is important because it takes us from the speculations of a sixteenth-century mystic to the modern scientific view that we are fortunate to enjoy today. But the story is also important because we can project its likely course into the future to show us where exoplanet research is heading and what questions we can hope to answer in the coming decades. For there is one question that has beguiled the minds of astronomers and poets from the Renaissance age of Giordano Bruno right up to the present: Are the planets inhabited? We are not the first generation that has imagined itself on the cusp of being able to answer that question, and our story of exoplanets will ultimately have to consider whether we are looking today at the right planets, with the right technology, and, perhaps of greatest importance, asking the right questions of them.

It turns out that Giordano Bruno's words would prove to be deeply prophetic for modern astronomers because, with only a couple of notable exceptions, exoplanets today remain masked by the glare of the stars they are bound to in orbit. Yet unseen does not mean undetected. It is an ironic shift of perspective to realize that, for all that we talk about the discovery of new exoplanets and measure their properties, in most cases we achieve this without ever having seen the planet directly. Most of the planet-hunting techniques that astronomers have in their arsenal are *indirect*, which means that they detect the *effect* of the planet on its host star rather than the planet itself. And it is the gravitational effect of a planet on its stellar parent that concerns our story today.

The example of our own solar system can help us understand the apparent paradox of indirect planet detection. Jupiter is the dominant sibling in our family of planets, and it follows an elliptical orbit about the sun, bound to its path by the force of the sun's gravity. Yet there can be no effect without consequence, and as the sun exerts a gravitational force on Jupiter, so must Jupiter in turn exert an equal and opposite force on the sun. But what effect could Jupiter possibly have on the sun? Jupiter takes twelve years to orbit the sun, and as it does so it tugs the sun around a similar but much smaller orbit of its own. To get a picture of how this strange back-and-forth motion might look, we can take a moment to imagine an alien astronomer observing us from across the Milky Way. If their telescopes were similar to ours, they would probably be able to detect our sun, but they might be hard-pressed to make out the light coming from any of the planets of our solar system. Yet from their perspective, that tiny tugging motion of the sun would give them a critical clue pointing toward the gravitational effect of unseen planets. The alien astronomers would see the sun complete its miniature orbit every twelve years, and if they did their math correctly, their equations would reveal the planet Jupiter with as much certainty as if they had observed it directly.

Far from having to invent alien astronomers to see this idea in action, we humans have already used it to discover unseen planets in our own solar system. In 1846 the French astronomer Urbain Le Verrier was attempting to understand why the orbit of Uranus—then the outermost known planet of the solar system—deviated from the orbit predicted by Newton's laws. The combined gravitational effect of the sun and the inner planets could not describe the motion of Uranus, so Le Verrier predicted the existence of a new, unknown planet beyond it. So precise was his mathematical analysis that, on September 18, 1846, he wrote to Johann Galle, then director of the Berlin Observatory, giving him the predicted position in the sky of this hidden member of the solar system. Fortune indeed favored the precise, as five days later Galle and his colleagues discovered a new planet, Neptune—at that point still dim and mysterious—just one degree on the sky away (about two full moons in width) from the position Le Verrier predicted.

The lessons drawn from our own solar system teach us the need for two qualities when searching for exoplanets—patience and precision. Patience is required to follow an alien sun through twelve years and more of motion in the hunt for the tiny tug of a Jupiter-like planet. Yet patience will gain you nothing without the precision to be able to detect the small velocity shift of the star as it wobbles back and forth in space. Jupiter imparts a velocity of thirteen meters per second to the sun, about the speed of the fastest sprint athletes. In contrast to this, our Earth imparts to the sun a velocity of just ten centimeters per second—a slow shuffle compared to the pace of a hundred-yard dash. For a new generation of terrestrial astronomers seeking exoplanets of their own, the challenge would be to build instruments with the required sensitivity to detect those tiny wiggles in the motion of their host stars.

It would ultimately take 150 years from Galle and Le Verrier's discovery of the last giant planet in our solar system to the discovery of the first Jupiter-like world in orbit about a star other than our own sun. In order to follow that story we will pick out a single golden thread of innovation that shines through the many interwoven ideas that led to the discovery of exoplanets. For it is the invention of astronomical spectroscopy that would create the new science of astrophysics and would at last bring the stars, and the planets they host, within our reach.

It might surprise you to learn that it is relatively easy to observe the spectrum of a star—albeit a very nearby one. In fact anyone can do so with a little preparation. It was Isaac Newton who showed the way by allowing a narrow shaft of sunlight passing through a gap in his curtains to fall upon a prism on his desk. The prism dispersed the light into a rainbow band of colors and revealed the spectrum of wavelengths that make up sunlight, the light from our local star. It was a revelatory observation that formed just a tiny part of Newton's monumental research into the properties of light—his *Opticks*—published in 1704.

Newton's insight would prove to be the first step along a path of discovery that would ultimately lead to a greater knowledge not just of the sun's nature but of the stars themselves. Throughout the nineteenth century scientists such as William Wollaston and Joseph von Fraunhofer worked to disperse sunlight at greater resolution—essentially into a broader band of color—and in doing so they noted the presence of dark, narrow lines superimposed upon the

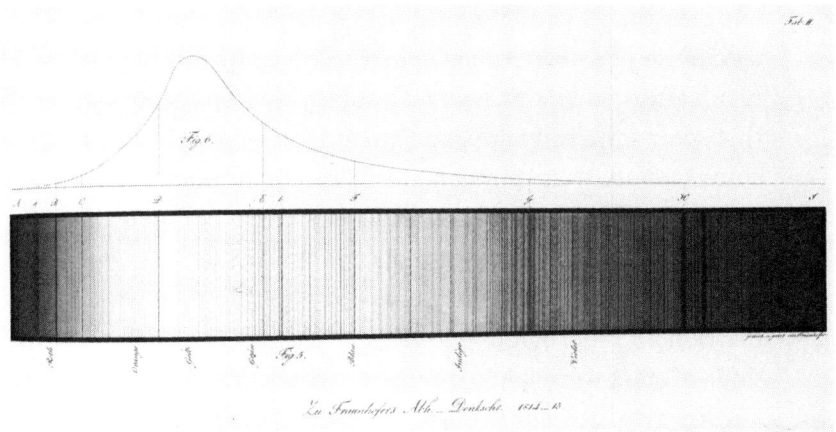

Fraunhofer's solar spectrum An engraving from Joseph von Fraunhofer's 1814–15 paper describing his observations of the solar spectrum. The graph in the top panel indicates the overall brightness of the sun at each wavelength, with redder light on the left and bluer on the right. Yellow light, corresponding to the wavelength where the sun is brightest, lies where "Fig. 6" is marked on the upper curve. The dark bands in the lower panel, some marked by letters above them, are the famous Fraunhofer lines that are caused by atoms of particular chemical elements absorbing light from the sun. Credit: Biodiversity Heritage Library.

bright spectrum of the sun. Eventually Gustav Kirchhoff and Robert Bunsen, beloved of science students of all ages, realized that these dark bands marked the precise wavelengths at which common chemical elements in gaseous form would absorb light. The inference was that the surface of the sun—the photosphere—must be a hot, tenuous gas that contained many of the same elements.

The invention of photography midway through the nineteenth century and its application to astronomy completed the technical revolution. When placed at the end of telescope, a spectrograph—the name we give to the device for dispersing light into its constituent wavelengths—would render the spectrum of a star onto a photographic plate. The images it captured could then be examined microscopically and quantitatively, an innovation that helped create the new discipline of astronomical spectroscopy. The intellectual reach afforded by this new technology was vast—for once this seems the only appropriate adjective. Astronomers could now study the physical conditions in the photosphere of any star for which their telescopes could collect sufficient photons. The science of astronomy grew to encompass a completely new discipline—astrophysics—in which the rich and incredible phenomena exhibited by populations of stars were laid open to human scrutiny.

Our own story of the path to exoplanets moves on to 1890, when a pair of American astronomers, Antonia Maury and Edward Pickering, working at the Harvard College Observatory used the new science of stellar spectroscopy to observe the curious motion of a previously unknown binary star. They were analyzing a series of spectra of what was thought to be a single star—Mizar A, itself twinned with Mizar B in the handle of the Big Dipper—that had been taken over multiple nights of observation. Maury noted that a particular dark line of calcium in the star's spectrum would periodically broaden and resolve itself into a close pair of lines before merging together again. The mathematics describing how the velocity of a star could shift the absorption lines in its spectrum had been developed in 1842 by the Austrian scientist Christian Doppler, and with the advent of spectroscopy astronomers had used the "Doppler effect" to measure the velocities of individual stars as they traveled through our Milky Way. However, Maury and Pickering correctly concluded that what they originally thought was a single star was in reality a binary pair, and that the two dark spectral lines arose from separate, rapidly moving stars in an orbit too close to resolve visually through their telescope. With the discovery of the first spectroscopic binary star, Maury and Pickering had revealed the dance of distant suns, but, more importantly, their technique offered a route by which others might one day reveal strange new worlds in orbit about them. For if we could now detect the velocity change of one star in orbit about another, in the future might improved telescopes and spectrometers allow a new generation of astronomers to detect the motion of a planet about its parent?

The next fifty years would see the rise of astrophysics, as stellar spectroscopy helped reveal the surface properties of stars and in turn hinted at their inner physics. Yet it was the years that followed World War II that witnessed not only the birth of the space age but also a technological revolution in astronomy. New and larger telescopes equipped with more capable instruments spurred the need for ambitious new plans—novel observing campaigns directed to answering important new questions in astronomy. Following the 1948 commissioning of the giant two-hundred-inch Hale telescope at Palomar Observatory, astronomer Otto Struve, an expert in the astrophysics of stars, laid out a plan titled "Proposal for a Project of High-Precision Stellar Radial Velocity Work." Struve realized that new telescopes and spectrographs would steadily improve the ability to measure Doppler shifts in the spectra of stars. That technological journey would begin with an astronomer of the early 1950s able to measure a velocity for a star moving back and forth in space as small as one kilometer per second (km/s)—a precision sufficient to discover the gravitational tug of a faint, low-mass star in orbit about it. Yet by the mid-1980s that same astronomer would now be able to reliably measure a velocity as small as ten meters per second (m/s)—a precision that would enable them, whether they realized it or not, to detect the influence of a Jupiter-sized planet about its parent star.

Struve, however, made a critical point in his paper that would be overlooked by later generations of astronomers, who then missed a vital shortcut on the path to detecting exoplanets. Struve's observation was deceptively simple: We observe stars orbiting other stars in very small orbits, as small as 0.05 astronomical units, or one-twentieth the distance from the Earth to our sun, so why should we not observe planets moving in similarly tight orbits? Depending on the mass of the stars in such binary orbits, they can complete one corevolution in as little as a few days. An astronomer searching for the varying Doppler signal of a star orbited by a planet at such a small distance could—with the required velocity sensitivity—confirm the reflex motion of the star in just a few nights of observation instead of the years or even decades required to confirm a planet in a twelve-year, Jupiter-like orbit.

Rather like a hiker who is perhaps too focused on an old and outdated map to notice a fork in the trail, astronomers missed Struve's signpost showing an easier route to Mount Exoplanet. In part the suggestion was overlooked because Struve's conception of what a planetary system *could* look like didn't match what *our* solar system actually did look like. The conviction that our home system appeared the way it did because that was just how nature produced planetary systems ultimately caused astronomers to miss a vital opportunity. However, perhaps this oversight instead arose from distraction, as, at the same time that Struve was setting out his plan, attention was diverted by the academic furor that accompanied the claimed detection of planets made by astronomers from a small but respected college observatory during and following the years of World War II.

In 1943 Danish astronomer Kaj Strand, then a research associate at Sproul Observatory, attached to Swarthmore College in Pennsylvania, claimed to have detected an anomalous motion in the dance of the nearby binary star 61 Cygnus across the sky. Strand's astrometric method of searching for planets involved a subtle shift of perspective on the part of the planet hunter compared to the radial velocity method we have so far considered. Imagine that you are watching an evening of ballroom dancing, with the slightly strange caveat that you can see only one partner in each waltzing couple. The perspective afforded by the stellar radial velocity technique is akin to standing at the edge of the dance floor and noting the subtle motion as the apparently unaccompanied dancers swing either toward or away from you. In contrast, the astrometric method is like looking straight down on the dancers from a balcony. In this case, no one is moving toward or away from you, and the unseen partner of each dancer whirls the other in subtle spirals as they gyrate around the room.

Strand attributed the additional wobble he detected in the 61 Cygni system to the gravitational pull of an unseen Jupiter-mass planet in a five-year orbit about the stellar pair. Then in 1963 his boss at Sproul, Peter van de Kamp, claimed to have detected a Jupiter-like planet in orbit about Barnard's star, just eleven light years distant, after twenty-six years of diligent monitoring. The astronomers of Sproul Observatory certainly seemed to be onto something that eluded the wider astronomical community, since no other astronomers were able to duplicate their findings. In time, however, it was van de Kamp's successor as director of Sproul Observatory, the German astronomer Wulff Heintz, who systematically deconstructed the evidence for each claimed planet by showing that all the stars in the field of each photographic plate, not just those claimed to have a planet, showed anomalous motions. Wulff's reanalysis of the Sproul plates revealed flaws in the way the photographs themselves were captured, and at a stroke he refuted the previous claims of planetary priority. Peter van de Kamp, who had garnered considerable publicity from his claims, never forgave Heintz for what he considered to be an act of profound academic disloyalty. As the curtain falls on this brief but significant scene, one can only imagine the sense of arctic chill that must have pervaded the corridors of the observatory offices . . .

This was the scientific landscape into which the first truly modern searches for exoplanets emerged. Otto Struve had challenged the astronomical community with his idea of measuring stellar motions with greater precision than ever before, yet there remained the risk that any astronomer foolhardy enough to claim the detection of planets about those stars would face the ridicule of their professional colleagues.

Key to achieving Struve's goals would be increasing the precision with which a star's radial velocity could be measured. Larger telescopes certainly capture more light, but that innovation alone would not be enough to reach for the stars and the planets that accompany them. The biggest jump in sensitivity was brought about by the replacement of photographic plates with digital detectors, in this case the early forerunners of video cameras. Not only were such detectors more effective than photographs at capturing photons, but the digital electronic record they produced could be analyzed relatively easily using early generations of computers.

The second innovation was much more subtle but no

less important. To advance from a radial velocity precision of 1 km/s in 1950 to 10 m/s in 1985—a factor of 100 increase in precision—astronomers had to address the very stability of the spectrograph itself. Changes in atmospheric pressure, even individual wind gusts, can vary the air pressure in a spectrograph and with it the refractive index along the path taken by the light beam. If left uncorrected, this effect would see the entire spectrum wobble back and forth on the detector, with a consequent smearing of wavelength information. One solution to this problem is to enclose the entire spectrograph in a pressure chamber to isolate the instrument from all atmospheric changes. The other is to introduce your own spectral absorption pattern into the light beam from the star. Whenever any change affects the light from the star, it will also affect your reference pattern as well. The simplest way to achieve this is to place a sealed transparent flask containing a low-density gas in the light beam from the star. Absorption lines from the gas appear as dark bands on the stellar spectrum and provide a built-in reference to which the star's own absorption lines can be compared.

This was the idea astronomers Gordon Walker and Bruce Campbell pioneered in the late 1970s. The gas cell they used produced a forest of narrow absorption lines that sat on top of the stellar spectrum yet avoided obscuring key absorption lines in the star used to measure its Doppler shift. When the resulting spectrum is viewed, the effect looks very much like a band of stellar color viewed through the teeth of a comb, and an alternative name for the technique they developed is that of a "spectral comb."

Using gas-cell spectroscopy, Walker, Campbell, and Stephenson Yang were among the first researchers to monitor the Doppler shift of large samples of stars in search of exoplanets. They sought the signal from a Jupiter-like planet in a multiyear orbit about a sunlike star, and the search consumed years of their careers. Like any explorers on unknown seas, there were the siren calls of false positives waiting to lead them astray. Among the most common of these were star spots, which vary the absorption properties of a star's photosphere and can mimic a Doppler signal. The closest they came to detecting a planet is described in their 1988 paper "A Search for Substellar Companions to Solar-Type Stars." Most of the stars they looked at showed no detectable velocity variation, yet one, HR 8974—better known as Gamma Cephei—showed a periodic shift in its velocity that amounted to 25 m/s repeated every 2.7 years.

So when I say that they came close to detection, that is not quite correct—they *did* detect an exoplanet in Gamma Cephei, in this case with a mass 1.7 times the mass of Jupiter. However, the team was wary of promoting its detection as the discovery of the first exoplanet. This reluctance was not because they didn't want to believe it. Far from it. Yet Walker, Campbell, and Yang were acutely aware of the ghosts of planets past, and as careful scientists they knew of the tricky effects in stellar photospheres that could mimic their result. Many in the present-day exoplanet community rightly recognize the significance of Walker, Campbell, and Yang's result and laud their contribution to the field. But—as we shall see—the Nobel Prize would eventually go to another team.

With hindsight we can see a scientific discipline poised on the threshold of discovery, not so much hesitant to take the next step as awaiting a smoking gun—a truly clear detection that would rise above the skepticism of the wider astronomical community. The path to discovery in the search for exoplanets contains many examples of false claims in to addition to those that have been proved real yet were hedged at the time. The long-sought smoking gun that revealed the existence of exoplanets was finally discovered in 1992—spectacularly so—yet once again we would confront the issue of the "wrong kind" of planet and ask if these really were the kind of worlds we had been looking for all along.

In any good horror film the audience is held in the grip of growing suspense before being shocked by an unexpected reveal—the curtain pulled aside, the flash of a clawed hand in the darkness. However, when two, then three of what would be referred to as "zombie" planets were revealed in orbit about a distant pulsar, the audience reaction was decidedly muted.

A pulsar is a rapidly rotating neutron star that produces a blip of radio emission once per revolution as viewed by astronomers on Earth. In this case PSR 1257+12 pulsed as it revolved every 6.2 milliseconds—that's 160 times a second, or about six times as fast as your average bullet blender. Given that the neutron star itself is only about ten kilometers across yet is more massive than our sun, it must present an awesome sight.

When viewed from the giant Arecibo radio telescope in Puerto Rico, however, PSR 1257+12 showed a subtle yet regular anomaly in the time when each radio pulse was received. Sometimes the pulse would arrive a fraction of a millisecond earlier than expected, sometimes it would arrive a fraction later—a pattern consistent with the reflex motion imparted by not just one but three planets in orbit about it. If that wasn't enough, then the properties of these planets would turn out to be equally astonishing. The radio timing results were incredibly precise and had revealed the existence of two planets of masses approximately equal to four Earth masses apiece and one tiny planet just two times as massive as Earth's moon. All three planets were orbiting their parent pulsar within the orbit of Mercury in our own solar system.

It was and remains a staggering discovery. A dim, dead star, lit not by the glowing fire of nuclear fusion but instead by a tenebrous revenant limned in the nebular glow of ionized gas trapped within its crushing gravitational environs. It is not clear whether the planets themselves are the blasted relics of the worlds that existed before the once-living star exploded in a supernova or whether they represent the very ashes of that explosion that accreted to re-form the undead planets that we observe today.

Astronomers Aleksander Wolszczan and Dale Frail announced the discovery of the first two planets of the PSR 1257+12 system in 1992. The third, moonlike world was revealed two years later. The discovery made a splash, although it was not as large as you might have expected. The field of astrobiology likes to talk about exoplanets orbiting "normal" stars. In this case "normal" is taken to mean a main sequence star that produces radiant energy by thermonuclear fusion. To many astronomers and astrobiologists, PSR 1257+12 was definitely not a "normal" star, and in the search for the first "true" exoplanet it just didn't count. To be fair, from an astrobiological perspective, you can see why. The planets of PSR 1257+12 would be dark, cold worlds whose exposed surfaces would be constantly swept by lethal winds of high-energy particles. From the perspective of Goldilocks, the planetary companions of PSR 1257+12 were definitely "too weird" to be of interest in the search for life. Thankfully, though, for a scientific field impatient to be going places, astronomers would not have to wait much longer to finally discover if not exactly the kind of planet they had been searching for all along, then at least something much closer to the mark.

The exoplanet 51 Pegasi b was presented to the world by Michel Mayor and Didier Queloz at an astronomical conference on October 6, 1995, in Florence, Italy. The following month the discovery was published in the prestigious journal *Nature*. It was everything astronomers had been looking for: a Jupiter-mass planet imparting a textbook reflex motion to an almost exactly sunlike star. The Doppler velocity signal measured for the star 51 Pegasus was slightly larger than that observed for Gamma Cephei, but the key characteristic of its accompanying planet, 51 Peg b, was its incredibly short, four-day orbit that placed it just 0.05 AU (astronomical units) from its parent star. At that distance—eight times closer to the sun than Mercury—this Jupiter-mass world would experience upper atmosphere temperatures of more than 1,400°C.

When the discovery of 51 Peg b was first announced, the scientific community remained skeptical of the claimed exoplanet. The orbit seemed far too short to be believable, at least to everyone except the late Otto Struve, and one technical criticism leveled at Queloz and Mayor was how they managed to compress multiple years of monitoring data into a single 4.2-day period. However, following the announcement, astronomers Geoff Marcy and Paul Butler, who had made many of their own important contributions to the search for exoplanets, used their upcoming October visit to Lick Observatory in California to confirm the existence of 51 Peg b by observing it over a single four-night period. It was this critical observation that, in the minds of many scientists, helped push 51 Peg b over that all-important threshold of community acceptance.

Exoplanets had finally arrived, and, although Goldilocks may well have been tempted to think 51 Peg b was far too hot to be of interest in the search for life, a key scientific breakthrough had nonetheless been made. In 1996 astronomers revealed six more exoplanetary systems using the radial velocity technique—including the first triple planetary system (after PSR 1257+12, of course) in Upsilon Andromeda. Since that time the number of exoplanets detected by all methods has roughly doubled every twenty-seven months. As of 2024, we know of some 5,606.

Of the thousands of known exoplanets, numerous record breakers stand out. The hottest world is KELT 9b, and at just under 4,500°C it's the very hottest of hot Jupiters. The most diminutive is Kepler 37b. It is only slightly larger than our own moon, yet with a surface temperature of at least 450°C it would likely be as welcoming as a hot oven. There's even an exoplanet that most closely resembles the planet Tatooine in *Star Wars*. Kepler 16b orbits twin suns and likely experiences the same kind of evocative sunsets that Luke Skywalker would wistfully gaze upon (though, unfortunately, Kepler 16b itself is not a dry, desert world but a Jupiter-mass planet in a Venus-like orbit). Each of these superlative planets, and there are a great many of them, is a twinkling gem we can't help but be drawn to, each gleaming with facets of possibility that beckon to our ceaseless curiosity.

The discovery of exoplanets was a crowning glory for astronomy. One of the great questions we had long asked of the cosmos had been answered, and the four-hundred-year dream of discovering planets around other stars was realized. Today we can look back some thirty years since the discovery of exoplanets, and in that time the fraction of stars that host planets has now been measured as a result of systematic and dedicated effort. The value of that number lies between 0.5 and 1 and describes a universe that is, quite frankly, astonishing. It means that when you look at each and every star in the night sky, there is at the very least an even chance that it is accompanied by an unseen planet. That same night sky you stand beneath today looks no different to our eyes than it looked to the generations of sky watchers for whom planets were just a dream, and yet for us it is profoundly changed, sown as it is with new cosmic possibilities.

For the first time since the Drake equation was written down in 1961, a previously unknown term on the road to life had been revealed. However, at the start of this chapter we recognized that one question concerning exoplanets trumps all others: *Are they inhabited?* Having discovered exoplanets and gone some way to determining the fraction of stars that host planets, the Drake equation tells us that the next logical question to answer is what fraction of these planets host life. However, it is also more than that. Whether life exists on these planets could be considered *the* question, the only one that really matters—at least to Earth-bound astrobiologists considering the question of life in the universe from the wrong side of discovery.

Perhaps not surprisingly, the question of whether exoplanets are inhabited has so far proven to be an ambitious one to answer, and astronomers have had to content themselves with asking a much simpler question: Which of these planets might be *habitable*? The distinction is important, because what astronomers actually mean by the word *habitable* are planets that might present conditions that are comparable to Earth, where conditions for life as we assume it to exist could at the very minimum be *tolerated*. When push comes to shove, our pale blue data point still remains the one to which all others are compared. However, the *tolerable* zone where planets might exist about their parent stars has less of a ring to it than the *habitable* zone where life might thrive and, for better or for worse, we use the idea of the habitable zone when sifting through the growing catalog of known exoplanets for those that, at the very least, *might* host life.

In my introduction to this chapter we passed a number of telescopes perched along the rocky ridge of La Silla Observatory. A discovery made by one of these telescopes is particularly useful to us because it provides a clear example of the search for habitable exoplanets that followed the confirmation of 51 Peg b in 1995. I highlight it for the additional reason that I am always keen to celebrate the scientific achievements of smaller telescopes, as they provide a necessary counterpoint to the preconception that it always takes big telescopes to make new discoveries. It is important to remember that stable spectrographs and imaging cameras operating on smaller but dedicated telescopes can still provide us with new and unexpected views of the wonder of exoplanets.

The tallest of the bright white domes at La Silla Observatory is that of the 3.6-meter telescope. It is some fifteen stories high and dominates the farthest and highest crags on the ridge. As you drive along the access road, your eye is naturally drawn to its gleaming form, and you might easily miss the old Swiss T70 telescope, the dome of which is barely larger than that of a local community observatory. The mirror of the Transiting Planets and Planetesimals Small Telescope (TRAPPIST) is only sixty centimeters in diameter, giving it a collecting area thirty-six times smaller than the mirror of the 3.6-meter telescope. However, the properties of this diminutive telescope and its first and most famous confirmed exoplanetary system reveal that, in the planet-hunting game, bigger isn't always better.

In what turned out to be a bumper year for planet hunting at La Silla, 2016 saw the announcement that the TRAPPIST survey had discovered its first planetary system. The team had been monitoring a dim red dwarf star using the transit method to detect the regular, planet-sized dip in the light received from the star as an unseen planet passed in front of it. The star itself, now known as TRAPPIST-1, is utterly puny, and we can detect it only because it is so close to Earth. At forty light years distant, it is a relatively close neighbor in our local stellar suburb. With a mass only 9 percent that of our own sun, it shines two thousand times less brightly and is barely massive enough to strike the flame of nuclear fusion in its core. The radius of this tiny star is only one-tenth that of the sun, meaning its stellar disk is smaller by a factor of 100. That small area turns out to be critical for planet detection, as even an Earth-sized world can produce a detectable fractional dip in the light received from the star. And that is what the TRAPPIST team detected—though in this case it would turn out to be not one but a staggering seven Earth-sized worlds orbiting this one star.

Once astronomers had discovered seven worlds in orbit about a single star, the fairy-tale metaphors they used to describe them became horribly confused. Tradition dictates that some suitable Snow White metaphor should accompany these seven planets in orbit about a single star but, for better or for worse, it is Goldilocks's name that astrobiologists have associated with the key question of

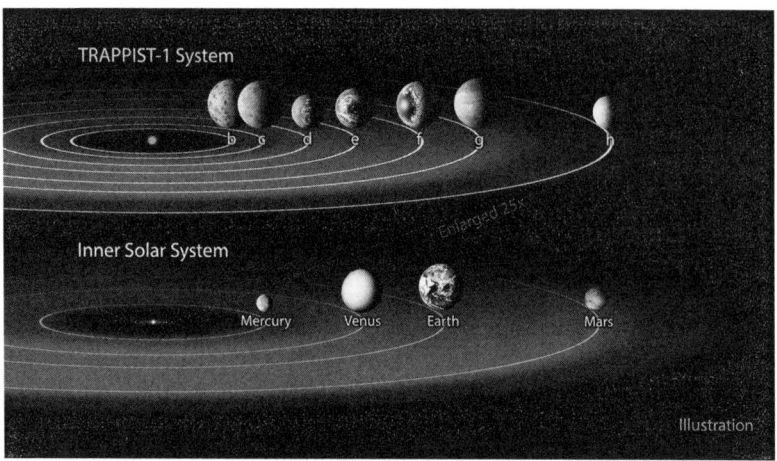

TRAPPIST-1 Worlds in miniature. The seven Earth-sized planets of the TRAPPIST-1 system orbit a dim red star some forty light years from our sun. Yet if we were to superimpose the TRAPPIST-1 planetary system on our own solar system, all seven worlds would lie well within the orbit of Mercury. The closest planet orbits its sun in just 1.5 days; the farthest takes a little under nineteen days. Diminutive though the system is, planets e, f, and g exist within the "habitable zone" about their star and may thus maintain planetary conditions suitable for life. Credit: NASA-JPL/Caltech.

habitability: What planetary conditions give rise to a world just right for life?

Leaving the mathematics to one side, we can all appreciate some of the factors involved in answering this question. Just like a campfire, a star is a source of radiant energy. Stand too close to the burning glow, and you will start to bake. Step back from the fire, and the intensity of radiant energy will drop off to the point where you start to feel cold against the chill of the night. Thankfully you can choose the best distance from which to enjoy the fire, standing either closer or farther away as it suits you. A planet in orbit about its star is not so fortunate. Once a planet or system of planets settles down after its formation, with rare encounters excepted it will retain its ordered array of orbits for billions of years.

The principle that underlies most if not all ideas of planetary habitability centers on whether surface conditions can support the stable existence of liquid water. The existence of water itself, liquid or not, is typically taken for granted—H_2O is a common molecule in the universe and has been observed in environments from prestellar gas clouds to the few glimpses of exoplanet atmospheres we have thus far obtained. Yet for a hot planet too close to its parent star, that water may well exist exclusively as vapor in its steamy atmosphere. For a cold planet, shivering on the periphery of a star's radiant glow, that water will freeze to form a surface sheen of ice. The range of orbital distances from the star that are just right for the existence of liquid water—the beloved green band depicted in press release images of exoplanetary systems—is the basis of what astronomers call the habitable zone.

Interestingly, if you worked through the math for our own solar system, then planet Earth would fail this habitability test. If we limit ourselves to including just the effects of solar heating and the re-radiation of some of that energy back into space, Earth would achieve an average global temperature of approximately minus 21°C. However, our pale blue data point shows us that the biggest factor to be taken into account after solar heating is the warming effect of the planet's atmosphere. Like a warm blanket, Earth's atmosphere retains infrared radiation that would otherwise be lost into space, which provides an extra forty degrees to our planetary heat budget. That extra heat is just enough to take us from a freezing ice world to a balmy planetary average temperature of 19°C.

All of these calculations, however, are filled with nuance

and assumption. The effect of an atmosphere can work both ways, either warming a cold planet, shivering just beyond the far edge of the habitable zone, or, if shielded with clouds, providing some blessed relief from stellar heating on the inner, hotter, edge. In addition, these calculations represent global average temperatures across a planet's illuminated disk. Just as we find on our planet today, the poles of our exoplanet would be colder than its average temperature, and the equator would be hotter. This means that the ideas the Goldilocks zone are based on are more complicated than one might think, being defined not just by a planet's distance from its star and by its atmosphere but also by the latitude on the planet itself. You can, of course, account for this by widening the soft green band of habitable orbital radii to account for localized holdouts of liquid water on planets close to the edges of the habitable zone. Yet, taking all of these ideas together, it is clear that the boundaries of the habitable zone are blurred and tenuous, a reflection of our lingering uncertainty as to the true properties of any particular exoplanet.

Within the TRAPPIST-1 system, three of the seven planets lie within the habitable zone of their dim red dwarf star, and the system as a whole provides a wonderful miniature portrait of how the idea of habitability is actually used. Saying that three of the seven planets are habitable does not mean that the prospects for life on the other four are zero—one could imagine life existing beneath the ice sheets of frigid, Europa-like worlds. Nor does it mean that the chances for life on the three so-called habitable worlds are likely to be high—we just don't know. The real utility—and limits—of the idea of habitability is that it allows astronomers and astrobiologists to *prioritize* for closer inspection those worlds that offer the best chance for hosting life.

This idea is important because at this point in our story we find ourselves faced with the question of how to determine whether life is present on the strange and bewildering array of worlds that we have discovered so far. It is unlikely that we will be able to look in penetrating detail at each and every exoplanet we currently know, and focusing on habitable worlds is currently our best approach to deciding which planets deserve our closer attention. Yet even if habitable worlds offer the best chances of hosting life, what scientific observations must we perform to confirm its presence? Earlier in our scientific journey we were introduced to the idea of an atmospheric biomarker—the chemical signature of life imprinted in the gases present in a planet's atmosphere—and how it might be used to identify a living planet. This is the key idea that may unlock the secrets of life on exoplanets; for just as the dark bands of absorption in a stellar spectrum reveal the gases that make up its outer layers, so might the spectroscopy of an exoplanet's atmosphere reveal the gaseous clues of any life it contains.

Atmospheric biomarkers arise from the planetary-scale exhalations of all life on an exoplanet—they offer a global test of life in the form of "I metabolize, therefore I am." The ideal example envisioned by astronomers is to take a spectrum of an exoworld that looks just like home—a healthy proportion of molecular oxygen with just a dash of methane for flavor. On our own world these gases, one highly reactive, the other fragile, would not exist without the metabolic output of life on Earth to sustain them.

However, we have already been warned once during our story of exoplanets not to focus too much on worlds that look like our home system. What gases might Otto Struve point us to in our search for life? Perhaps he would counsel that we should learn on the job, that each exoplanet atmosphere we manage to capture in a spectrum will teach us, one world at a time, the rules of the chemical game that all planets must follow. Only then, perhaps when scrutinizing a habitable world, might we note some mix of atmospheric gases that bucks the trend, a planetary oddity where the usual rules don't seem to apply. Many scientists have speculated on what particular chemical elements and molecular species might unambiguously point to life, yet numerous others remain skeptical. How can one be sure that a particular molecule can be produced only by life when one's knowledge of the chemical rules of exoplanet atmospheres are still to be mapped out?

This is one aspect of the challenge we face in our quest to detect life on exoplanets. It is a deeply philosophical conundrum that goes to the heart of the questions of what life is and how we might detect it. Despite these challenges, the search for biomarkers in the spectra of exoplanet atmospheres is the route astronomers have selected as the best path to answering whether these brave new worlds host life. The second challenge we face is one of engineering and technology, and that is no less daunting. For just how do you obtain a spectrum of an exoplanet that may be lost in the glare of a parent star that is some millions to billions times as bright? To meet this challenge, astronomers will have to contemplate the construction of telescopes of truly vast size to capture the faint and elusive signatures of life on these planets. But mere size will not be enough. Ingenuity and innovation will have to advance in equal measure with telescope collecting power in order to take the next big step on our journey to detecting life on exoplanets. To see that future under construction, and to contemplate what it will achieve, we will have to leave La Silla Observatory and travel farther north along the Andes chain to visit another mountaintop observatory set in a land of giants.

I have spent a good fraction of my life at astronomical observatories, using telescopes both large and small across planet Earth, in addition to quite a few above it. Even as a seasoned astronomer, however, the new facility by the European Southern Observatory (ESO) under construction at Cerro Armazones in northern Chile fills me with awe.

The heart of the new telescope will consist of a mirror thirty-nine meters in diameter that it will use to view the cosmos. The collecting area of this colossal reflecting surface will be fifteen times larger than any competing observatory, either on Earth or beyond it. When William Parsons, the Third Earl of Rosse, constructed his revolutionary 1.8 meter reflecting telescope in 1845, it was dubbed the "Leviathan of Parsonstown." In comparison to that moniker, what name could possibly be suitable for a thirty-nine-meter monster?

ESO has named the project the ELT, the Extremely Large Telescope—though, to be fair, after naming their previous project the Very Large Telescope, there were not many options as to where to go next. The telescope along with its instruments and support structure will weigh some 3,900 tons, a mass that compares to the displacement of a modern US navy warship, yet the whole structure will be able to pivot across the entire sky to balance on a pinpoint of light. The dome structure to enclose this marvel will sit on a footprint equal to that of a professional soccer pitch, an entire stadium that can move in graceful partnership with the telescope as it tracks across the sky. The summit of Cerro Armazones itself was leveled in a gigantic explosion to create the site. Today, beneath the azure skies of northern Chile, the summit of Armazones is crowned with a huge concrete foundation from which the bare steel bones of the vast dome claw at the sky. The telescope is scheduled to see first light in 2028, and from its initial engineering conception it was always designed to be a discovery machine.

Lord Rosse's Leviathan was constructed in 1845 and was surpassed in size only in 1917 when George Ellery Hale constructed the hundred-inch (2.5-meter) Hooker telescope on Mount Wilson in California. That passage of seventy-two years brought an increase in collecting area of a little under a factor of two—albeit at a much better observing site than rural Ireland. Today one of the world's largest telescopes, at the W. M. Keck Observatory atop Maunakea in Hawaii, has a primary mirror ten meters in diameter, a sixteen-fold increase in collecting area compared to the hundred-inch when it was commissioned in 1993. The ELT, when completed, will collect just over fifteen times more light than Keck, 240 times more than the hundred-inch, and approximately five hundred times more than the Leviathan of Parsonstown. It represents a leap forward as great as any in the history of the telescope, and the single reason for building a telescope larger than any other is to collect more light than ever before. No matter what new questions we face—the most distant galaxies in the universe, the first stars to form in the cosmos, or the faint exhalations of life preserved in the atmosphere of an exoplanet—astronomers are seeking to answer them by collecting sufficient photons to reveal these most elusive of signals.

The leap of technology and light-collecting power represented by the Extremely Large Telescope is not a leap of faith, however. The ELT will take many of the technologies that we have already discussed in this chapter to levels that sound almost like science fiction. Throughout our journey together in this book I have presented the idea that our location on this pale blue data point offers several unique perspectives on whether life might exist beyond it. That idea finds its ultimate expression in the combination of the Extremely Large Telescope and instruments such as METIS—the Mid-infrared ELT Imager and Spectrograph.

Named after the Greek goddess of wisdom and cunning, METIS looks more like the *Apollo* lunar lander than an advanced astronomical camera. Encased within a low-temperature cryostat and supported on a stilt-like scaffold, this instrument deploys a suite of state-of-the-art astronomical ideas in the hunt for biomarkers on habitable worlds—an adaptive optics camera to still the ripples caused by light traveling through Earth's atmosphere and deliver superb images that will surpass those taken from space; a coronograph to occult the light from a parent star; and a high-resolution spectrograph capable of resolving a multitude of individual spectral lines in an exoplanet atmosphere.

What astronomers really crave are direct spectroscopic views of exoplanet atmospheres. But as we have learned, not just any old exoplanet will do. Habitable exoplanets are the key, and at present that means Earth-sized, rocky worlds orbiting within the habitable zones of stable stars. Just as the application of spectroscopy to the study of stellar atmospheres revealed the physical composition of stars, so too does exoplanet spectroscopy offer the promise of revealing their atmospheric composition and answering the all-important question of whether life plays a role in their physical stories.

The spectroscopy of habitable Earths with METIS is eminently achievable, though it would clearly be misleading to paint the project as an easy one. At the level of precision obtained by METIS, the spectrum of an exoplanet atmosphere is anything but static. The absorption lines from its atmosphere rhythmically Doppler shift in wavelength as the planet completes each orbit about its star. However, the known orbital elements of the exoplanet allow us to predict these shifts, and the features can be followed as they slowly cycle back and forth in the spectrum. Using this technique, either individual lines or important combinations of lines from potential biomarker molecules such as oxygen and methane can be tracked through each phase of the exoplanet orbit.

Whether observed directly using its powerful suite of adaptive optics or viewed as the planet transits its star, the Extremely Large Telescope will bring one overarching advantage to the field of exoplanet spectroscopy: its massive light-collecting power. Although there are many challenges along the path to revealing exoplanet atmospheres, the biggest hurdle of all is simply collecting enough photons to reveal the tenuous evidence of life they may contain. This is the evolutionary pressure in the lineage of telescope design that has given rise to an age of giants.

It is important to note, however, that European astronomers are not the only ones with their eyes on the prize of building the world's largest telescope. An international consortium of nations led by the United States is aiming to construct a Thirty Meter Telescope (TMT) on the summit of Maunakea on the Big Island of Hawaii. While slightly smaller than the ELT, the TMT would benefit from the higher site and access to world-renowned atmospheric conditions on the summit of this 4,200-meter volcano. However, in contrast to the ELT's construction in isolation on Cerro Armazones in Chile, the TMT represents a large addition to a mountaintop observatory already well populated with telescopes. Local opposition to further development on what many consider sacred ground stalled the TMT project for a number of years before the State of Hawaii intervened to reset the body tasked with stewardship of the Maunakea summit. The road map leading to TMT construction on Maunakea remains uncertain; yet if a path to reconciliation between the differing stakeholders can be agreed upon, a route to realizing this ambitious new observatory might yet be found.

This new generation of telescopic titans will scrutinize the handful of nearby habitable planets that we currently know. Will these few Earth-sized worlds be sufficient to answer the question of whether life exists on exoplanets, or will we have to cast the net wider to discover new worlds and improve the odds of finding life on any one of them? Although the next generation of telescopes and spectrographs represents a tremendous leap in capability, even their lengthy reach will only graze the very edge of what is required to detect and measure the detailed properties of exoplanet atmospheres.

The current crop of rocky worlds in orbit about dim red stars—the planets of the TRAPPIST-1 system plus a handful of others—represents the best chance for the next generation of astronomers and astrobiologists to determine whether life plays a role in the chemistry of their atmospheres. The reasons for this are twofold. First, the dimmer, cooler, and lower in mass the parent star is, the easier it is to make any measurement of a planet orbiting it. In every sense, there's simply less star to compete against. Second, and perhaps more subtly, given that these planets are among the closest and most habitable that we know of, almost every new telescope that aims to excel at exoplanet science will have been designed with these targets in mind. After all, although a bias remains for the "right" kind of planet—a true Earth twin in orbit about a sunlike star—the simple fact is that we are still not detecting such worlds in any numbers.

There is, of course, one pressing question: How many exoplanets will we need to observe in order to learn the chemical rules followed by their atmospheres? When faced with such nebulously phrased questions as these, astronomers, and perhaps scientists in general, often fall back on the "factor of ten" approach. If we can increase our knowledge by an order of magnitude, this may offer a reasonable, yet achievable, possibility of learning something new and important. To date, the just over five thousand known exoplanets have yielded a handful of Earth-like worlds. How will we discover fifty thousand exoplanets and bring the number of Earth-like worlds to several tens of examples? Two space missions that are currently flying and collecting data may well prove to be especially important in this wider hunt for new habitable worlds: TESS and Gaia.

NASA's current planet discovery mission, the Transiting Exoplanet Survey Satellite (TESS), is conducting an ambitious survey of the entire night sky. Launched in 2018 and still operating today, TESS is searching for short-period transiting exoplanets around some 200,000 of the brightest stars as viewed from Earth. It has so far discovered just under three hundred new exoplanets, with four thousand more candidates awaiting confirmation. Of these, some few tens of planets are expected to be super Earths (defined as being up to two Earth radii) orbiting within the habitable zone of an M-dwarf star. While that may sound like a limited haul, it represents that magic factor of ten increase that we previously hoped for—and it is happening right now.

TESS continuously observes a single patch of the heavens for twenty-seven days straight before moving on to the next swath of sky. With the need to detect at least two consecutive transits to confirm a system, this effectively limits TESS to detecting planets in less than thirteen-day day orbits. When we consider Kepler's laws of planetary motion, an orbital period of this duration would place a two-Earth-mass world 0.06 AU away from a typical M-dwarf star—smack in the middle of the star's habitable zone. The restriction to observe each patch of the sky for a limited time means that many potential worlds around the very same stars with confirmed exoplanets might exist yet would remain undetected by TESS. Any exoplanet that takes longer than twenty-seven days to complete one orbit might remain completely undetected. However, a small fraction of the sky at the poles of the celestial sphere has remained continuously visible to TESS during its survey of the heavens and will mitigate this effect.

With this limitation in mind, it's interesting to consider the discoveries that are anticipated from another current space mission—ESA's Gaia astrometric mission. Like TESS, Gaia is also conducting a survey of almost the entire night sky. But whereas TESS's primary mission is to complete one complete cycle of observations across the sky, Gaia, which was launched in 2013 and is expected to continue observing until at least 2025, has accumulated an impressive seventy scans of the night sky—monitoring the positions of approximately one billion stars—during its first five years of operation.

Gaia's primary mission is astrometry—to map the positions of over a billion stars within our Milky Way. Yet in monitoring the detailed positions of stars, Gaia will be able to use the same astrometric method that detected the fake worlds of the 1950s and 1960s. In contrast to those efforts, however, Gaia offers a very stable, long-duration space platform from which to observe the tiny yet regular wobbles in a star's position on the sky. Using data anticipated from the first ten years of Gaia observations, astronomers expect to discover upward of seventy thousand planets outside our solar system. It is worth taking a moment to appreciate just how staggering that number is. It represents an increase by over a factor of ten in the number of known exoplanets compared to today's figure, which is all the more astonishing since the majority of these will be face-on systems where our viewing location on Earth appears to look down upon the dance of planetary orbits. This is a perspective never afforded to us by the transit and radial velocity searches with which we are more familiar. Given that the size of the astrometric wobble imparted to the star increases with the mass of the planet that tugs it, it will come as no surprise to learn that most of the planets Gaia detects will be Jupiter-sized worlds orbiting a few astronomical units from their parent stars—not good candidates for habitability. However, with seventy thousand planets in total, the rarer planets, Earth-sized and closer to their parent stars, may yet start to add up.

Both TESS and Gaia are examples of the "small is beautiful" approach to space observatories. TESS has four cameras, each with a collecting area equivalent to a thirty-centimeter-diameter telescope, whereas Gaia has a rectangular mirror that corresponds to a circular aperture about fifty centimeters across. Yet it is not surprising that a much larger space telescope steals all of the headlines: the James Webb Space Telescope. From our perspective as astrobiologists in search of life, we must now ask whether the hype surrounding this hyperobservatory is justified.

The James Webb Space Telescope was launched on Christmas Day of 2021, a fitting gift for the American, European, and Canadian astronomers who had spent the best part of two decades designing and constructing the space observatory. James Webb's 6.5-meter gold-coated, ultracold mirror is providing a new generation of astronomers with unparalleled views of the cosmos, from the accretion disks around supermassive black holes to the plurality of exoplanets that we have encountered so far in this chapter.

Webb has already demonstrated its potential as an exoplanet observatory by studying the atmospheres of some of the more massive gaseous worlds that we currently know. In a sequence of observations presented in 2023, it revealed in silhouette the atmosphere of the transiting exoplanet K2-18b and detected hints of carbon dioxide and methane in its tenuous outer layers. Webb's observations of K2-18b, orbiting in the habitable zone of a pale red star over a hundred light years from Earth, are important because K2-18b belongs to the one class of planet for which our solar system offers no analogues. If the planets of our solar system were arranged in order of increasing mass, K2-18b would sit in the conspicuous gap between Earth and Uranus, and, as a result, we have no supersized pale blue dot to compare it with. Yet these super Earths or sub-Neptunes are among the most common class of planet in the Milky Way, and they present a firm reality check to the approach we have taken so far in this book, in which we seek to learn about the universe by studying analogues closer to home.

Yet although orbiting within the habitable zone of its parent star, K2-18b does not fit our current expectation of what a living world might look like. The most straightforward interpretation of the observations to date is that it is a lightweight Neptune or mini-Uranus rather than a heavy Earth. Although astrobiologists remain open to the potential of such worlds to surprise us—we know very little about them after all—the top planetary targets remain the less massive, rocky worlds that orbit in the Goldilocks zone.

What are the prospects for Webb to be the first to detect the signatures of life in the atmospheres of worlds such as those of the TRAPPIST-1 system? Although James Webb is an exquisite space observatory, its mirror is thirty-six times smaller than the Earth-based Extremely Large Telescope. In the race to collect photons, Webb is the tortoise, and the ELT is definitely the hare. Webb will require many tens or even hundreds of transit observations per planetary system to achieve the merest detection of an atmosphere in a habitable, Earth-like world. Fortunately, though, the European Ariane 5 rocket that lifted James Webb into its Earth-following orbit did so with such precision that Webb conserved enough onboard propellant to keep its trajectory in check for the next twenty years. This extended mission lifetime may encourage NASA to indeed dare mighty things. The space agency has previously demonstrated with the Hubble Space Telescope Deep Fields projects that it was willing to step outside the box of normal observing practice to achieve unique and spectacular results with its premier space observatory. The prospect of detecting an atmosphere on a habitable world may well tempt it to eventually undertake a similarly ambitious program with James Webb—only time will tell.

As we approach the end of our story of exoplanets, it can be sobering to reflect upon the incremental pace of research and the fickle nature of discovery. In retelling the story of past advances we can compress the axis of time and allow our narrative to skip lightly from one moment of discovery to the next. Yet ultimately this gives a misleading impression of the actual pace at which knowledge advances.

The approximately thirty-year period from the discovery of 51 Peg b in 1995 to the present day has revealed that exoplanets not only exist but do so in a bewildering variety of configurations. In terms of the Drake equation, the wonderful chain of ideas that links the birth of stars to communicating aliens, we have spent those thirty years answering just one question: What fraction of stars host a planet? We now know that the answer is very close to one. The plurality of worlds, once the nocturnal dream of Enlightenment philosophers, is now established fact. One can only hope that Giordano Bruno and others who dreamed of planets beyond number, wherever they may now reside, might permit themselves a smile of satisfaction.

It sounds so simple to write that the fraction of stars hosting a planet is close to unity, yet establishing that required an immense investment of effort and ingenuity on the part of a generation of scientists. However, it also represents a single yet profound step along the path of knowledge. What will the next step look like? For all that I have attempted to describe one particular direction of present-day exoplanet research, can we look beyond the current prospects of albeit dramatic new telescopes to imagine what some future moment of discovery might look like?

It might be that all important elements are in play. Perhaps the most promising worlds of the TRAPPIST-1 system do represent our best candidates for exploring the atmospheres of habitable planets. The European Extremely Large Telescope project, twinned with quite incredible instruments, may well provide the means with which to engineer the first realistic search for biomarkers on alien worlds.

I don't doubt that surprises are in store for the first astronomers to study the detailed properties of exoplanet atmospheres. Some of these surprises might be less welcome than we hope for. We have learned that the current crop of rocky, habitable worlds orbit low-mass, M-type stars. While detecting planets about these feeble stars can be more straightforward than against the glare of brighter parents, those very same M-type stars are notorious for displaying violent magnetic storms and stellar flares. Perhaps the conditions for life on these so-called habitable worlds are much more fickle and unpredictable than we currently realize. However, other surprises might be astonishing—all the more so for being unexpected. The James Webb Space Telescope has already revealed some of the atmospheric properties of enigmatic planets in the mass range between Earth and Uranus. Given that our solar system offers us few, if any, clues as to what such worlds will look like, it would be foolish in the extreme to rule out the most common type of planet in the galaxy as a potential home for life.

If hindsight has taught us one thing, it is that discoveries rarely play out in the manner that one originally expects— just ask 51 Peg b. Yet how wonderful might that scenario be? Perhaps the first worlds to be explored will present anomalous detections of molecules previously overlooked as biomarkers that will, in turn, point to new patterns of life among the stars. One can only hope that some present or future astronomer is presented with the quip of a lifetime when they finally turn to their colleagues and remark, "It's life, Jim, but not as we know it."

On many of the nights that I worked at La Silla Observatory I would relax after my night of telescope operations by taking the long route back to my dormitory. With a contented team of visiting astronomers in charge of the telescope, I could take my leave at around two o'clock in the morning and follow the winding footpath that led down the mountain from the summit.

The first few steps upon the trail would leave the telescopes behind me while the stars overhead formed a silent, glittering dome. Moonless nights were my favorite. A full moon in a dark sky appears very bright indeed, blinding the eye to many of the stars above. A bright moon would saturate the landscape, not to the illumination of day, but certainly sufficient to dispel that magical sense of mystery that comes from walking alone at night, the sense of proximity to something present yet unknown.

On the nights when the moon was new or below the horizon, the stars would have free play across the sky. The barest dusting of silver light would settle on the mountainside, good enough for someone familiar with the path but not much more than that. However, the insubstantial, barely perceptible light would provide a subtle tingle of kinship, of connection with the night sky, as each photon that my eye sensed came not from scattered or reflected light emitted by our own sun but from the luminous fires of suns far beyond our solar system. Under such skies on such nights, those stars would not feel so very distant any more.

Like many of the experiences I have described in this book, that sense of communion is in many ways an act of the imagination, though it is no less powerful or important for that. When we discover the first atmosphere of a potentially life-bearing exoplanet—perhaps detecting the tantalizing signatures of oxygen, water, or methane—we will be no closer to encountering that life directly. We might, through direct imaging, catch a green sheen upon the surface of a vegetated planet or the specular reflection of sunlight from the waters of an ocean world. Yet the discovery will remain remote, separated from us by gulfs as large as any that our imagination has ever attempted to cross. In analogy to my moonless wanderings under those very same stars, that discovery will be no less powerful or important for that physical separation. Our sense of the cosmos will be forever changed, and our wanderings beneath the distant stars will, from that moment on, be a little less lonely.

5
TO CATCH A FALLING STAR

METEORITES AND THE CLUES TO EARTH'S
ORIGIN AS A LIFE-BEARING PLANET

In the final chapter of his beautiful memoir *The Periodic Table*, the Italian author Primo Levi muses upon the long and interesting chemical history of a single atom of carbon. For Levi this individual atom begins its story encased within a molecule of calcium carbonate, itself forming the crystal structure of a limestone rock on the surface of Earth. Baked, breathed, photosynthesized, and imbibed, the atom concludes its journey enmeshed in the cellular structure of the writer's brain, ending with the flourish of his pencil as a single dark dot of graphite on a white page.

Yet, sitting on my desk in front of me as I write these words are further atoms of carbon. These atoms happen to be encased in the minerals that make up a meteorite that I recovered on the fringes of the Sahara desert. For it is the story of meteorites that I want to share with you in this chapter: how they fall to our pale blue data point and the chemical tales they tell about the birth of our solar system and, perhaps, of life itself.

Levi acknowledged that his atom of carbon had a long and interesting cosmic history; but, as was his right, he began the atom's story at the moment that seemed most propitious to him. To begin our stories of meteorites we can take inspiration from Levi and trace the history of a different carbon atom to a point much further back in time. From our perspective on planet Earth, this particular carbon atom is special because its story begins long ago and far from our own sun, in the vicinity of a relatively cool yet immense red giant star. In astronomical parlance we call such stars asymptotic giant branch, or AGB, stars. It is an evolved star, perhaps as massive as our sun is today, but certainly older. The core of the star is a hot yet inert mass of carbon and oxygen. Perched on top of the stellar core are dense shells of gas where hydrogen still fuses into helium and helium fuses into carbon.

Our particular carbon atom was forged from the sudden and violent collision of three helium nuclei in the searing furnaces of an AGB star. Waves of gaseous convection and pulsation washed this shipwrecked nucleus to the star's tenuous, sloshing outer atmosphere. These turbulent shells of gas heat the star's outer envelope, causing it to expand and cool, eventually giving the star its characteristic dim red color as viewed by astronomers. The periphery of the star remains hot, as high as 3,300°C, and at such temperatures microscopic grains of refractory minerals can condense out of the gas, in this case as a metallic rain of silicon carbide, corundum, olivine, and pyroxene crystallizing out of vast stellar clouds. It is here that our carbon atom creates an important and lasting partnership. It forms a tight chemical bond with a silicon atom—its close cousin on the periodic table—to form a molecule of silicon carbide.

These tiny grains are swept up in the pulsating winds

that erupt from AGB stars at the ends of their lives, and it is these winds that provide the vital link between what is quite literally dust from a distant sun and a meteorite that you can hold in your hand. Each speck of cosmic flotsam is blown from its parent star with about the same velocity that space probes such as *Voyager 2* and *New Horizons* have flown from our own solar system. At such speeds it would take a cosmic voyager half a million years or more to cover the light-years of distance that separate the stars. We are always taught that these immense distances would prove an insurmountable barrier to any practical human method of interstellar travel. Yet what is this span of time to a microscopic grain of silicon carbide? No more than the merest blink of an eye in the lifetime of a star or solar system.

Our atom of carbon rode this stellar wind across the gulfs of space to land upon the shore of a new solar system and bear witness to the throes of its creation. Our infant sun, dim and shrouded, lay within a swirling disk of gas and dust. Molecules of condensing gaseous matter and dust grains coalesced to form pebbles of ashy material that, in turn, began to collide with one another in the still fragile grip of their tiny gravitational fields. The first planetoids began to form—young and boisterous giants that careened through the glowing disk accumulating more protoplanetary material. Enduring within its grain of silicon carbide, our atom of carbon witnessed the birth of planets in our solar system yet remained relatively aloof from the planet-building spectacle. Perhaps it once fell upon the surface of an asteroid, but its haven would soon have been torn asunder in later collisions. Our atom remained part of a somber and ashen rock, half-formed and friable. Eons passed. A distant dot of a planet that was once a molten sphere of rock cooled and became a smoggy brown mystery. Only later did the skies clear and reveal a pale blue planet. Some 4.5 billion years later in the history of the solar system, on February 8, 1969, as reckoned by us Earthlings, our paths finally converged.

The Allende meteorite fell to Earth through the skies above the Mexican state of Chihuahua in the early hours of that February morning in 1969. Over two tons of prized material was eventually recovered from the fall, which was quickly recognized as being of a particularly rare and precious type. Carbonaceous chondrite meteorites are thought to be among the very earliest relics of our solar system. They are richly adorned with tiny mineral chondrules— glassy beads of rock encased within a matrix of ash that, at least relatively speaking, abounds in both water molecules and an organic smorgasbord of complex carbon-bearing compounds. When another carbonaceous chondrite meteorite fell to Earth at Murchison in Australia later that same year, the scientist dispatched to collect the material remarked at the pungent odor emanating from the rocks, redolent of the aromatic smells of classroom chemistry. It therefore turns out that not only can you reach out and touch objects that preserve the earliest moments of our solar system, you can smell them as well. Yet, while offering us reminders of our earliest lessons in chemistry, might such ancient rocks reveal previously unsuspected clues regarding the origin of our biology as well?

My own slice of the Allende meteorite unfortunately possesses no such odor. What it does contain, however, are tiny inclusions of calcium and aluminum, bright sparks of light against the dull backdrop of gray chondrules. Careful analysis of the proportions of rare radioactive isotopes in these calcium-aluminum inclusions places them among the very oldest components of our solar system yet

Allende meteorite My own slice of the carbonaceous chondrite meteorite named Allende that fell to Earth in February 1969. The ghostly white imprints are the outlines of chrondules, the glassy pebbles and grains that accreted to form the parent asteroid of this meteorite. The bright white nuggets are calcium-aluminum inclusions, the very first solid bits of our solar system to condense out of the primordial disk of gas that orbited our young sun. Credit: Sara Ellison.

discovered. These tiny flakes of metal are 4.568 billion years old, and they were some of the first chemical elements to condense out of the hot solar nebula as it cooled from its initial collapse.

It is within these inclusions that we finally greet our far-traveling atoms of carbon, held within individual grains of silicon carbide. The lasting bond between a silicon atom and our carbon atom has endured through the eons until today, when they lie preserved microscopically on my desk in front of me. Their bond tells the story of the dying moments of a giant red star, when dust grains condensed out from the gas that makes up its vast, tenuous atmosphere. From there they were caught in the silent stellar wind emanating from the star and, like the ash from a bonfire on a clear night, carried up to the waiting heavens. Within our own solar system, these pre-solar grains, interstellar interlopers from the cosmos, carry within them the isotopic fingerprints of at least seven different and long-dead stars that fused them and then bequeathed their chemical legacy to us. I find this connection deeply comforting; for, although from our perspective on this pale blue dot we float, seemingly adrift, on the swells of cosmic night, our home islands of the solar system are in fact washed by stellar waves from many distant shores.

The story of our carbon atom to the point where you are reading it now is only the prelude to the wider story of what meteorites can tell us of the history of our solar system and the possibility of life existing beyond our pale blue dot. We will learn how science has gone beyond the compelling yet scattered physical stories told to us by meteorites to prospect for rocks, pebbles, and dust grains directly from their parent asteroids. Finally, and perhaps most striking of all, we will discover how a new generation of Earth-based observatories have revealed the first moments of planetary creation in solar systems beyond our own, providing astonishingly direct views of cosmic events that the meteorites we collect here on Earth only dimly hint at.

There is a tangible pleasure to be experienced in holding a meteorite. In one's hand they look like ancient coals from some primordial fire, but in reality the exterior fusion crust is a product of their relatively recent fiery descent through Earth's atmosphere. Their surfaces can appear ablated, a smooth finish rippled by shallow indentations—regmaglypts—thumblike imprints from the incandescent touch of their atmospheric fireball. But the deeper thrill of holding a meteorite in your hand comes from the tactile connection to a relic of the solar system that may have remained unchanged since it formed just over 4.5 billion years ago. Each one has a particular physical story to tell. Each is a wanderer through the cosmos that, had he been able to, Herodotus, the great classical collector of stories, would have questioned and told us of their travels.

I keep a modest collection of meteorites on my office desk, a select anthology of stories that tells the history of our solar system. Some are weighty and intact; others have been cut in half to reveal their inner secrets. Still others, the rarer and more expensive examples, have been sliced into thin sections—less like the pages of a book and more akin to an early Daguerreotype, a primitive photograph captured from a forgotten age. Each specimen tells of a particular phase in the creation of our solar system.

NWA 6135 is an unromantic name for a wonderful piece of rock. It is an LL 3.8 chondrite—a metal-poor meteorite that displays a rich assortment of relatively well-preserved chondrules. Its unprocessed interior tells us that it was never buried deep within any asteroid parent body and it emerged from the crucible of planetary creation relatively unscathed by pressure and heat. Its feast of fused, millimeter-sized chondrules preserves the earliest history of planetary accretion in our solar system, when planets were mere pebbles and Earth was a distant dream.

My one-kilogram chunk of the famous Campo de Cielo (Field of Heaven) meteorite is an iron relic from a failed planetoid. The young planet that preceded it had grown massive enough in the first few millions of years in the history of the solar system for an iron-nickel core to differentiate from silicon- and carbon-rich minerals and settle to the center of the young planet. Left to its own devices, it could have become a youthful Venus or even an Earth. The nascent planet survived long enough for the iron core to cool over perhaps 100,000 years and preserve the Widmanstätten pattern of ordered iron-nickel crystals within its structure. Yet fate saw this planetoid, perhaps as large as 1,000 kilometers across, collide with some unknown body. Both planetesimals were torn asunder in this cataclysm, and though most of their material may well have accreted again to form new planets and asteroids, Campo de Cielo is one of the chunks that got away, finally falling to Earth some 4,500 years ago in what today is northern Argentina.

From one point of view meteorites tell an apparently straightforward story of the earliest growth of our solar system: Gas and dust particles initially stuck together through surface electrostatic charges—a stronger force than their feeble gravitational attraction—then grew through a cascade of weak collisions to form larger, more massive and gravitationally robust planetoids. Yet meteorites can still surprise us. I have on my desk a small piece of the Gujba meteorite, which fell in Nigeria in 1984, that contains a mixture of iron-nickel nuggets from a disrupted planetoid encased in a matrix of grainy, ash-like minerals indicative of the most primitive stages of planetary growth. This important meteorite tells us clearly that the earliest stages of planet formation in our solar system overlapped each other in a repeating cycle of growth, accretion, destruction, and growth.

However rich in meteorite samples my office desk may be, there's no substitute for the scientific thrill of searching for these rare rocks in the field. Yet, to make this dream a reality, I would have to learn where in the world one actually goes to look for meteorites and how to recognize them once I got there. Although meteorites fall to Earth at random, scattered equally across all parts of our planet, successful meteorite hunters know to look for them in the locations where they are most easily found—in landscapes bare of vegetation where the local weathering of rocks looks subtly different from the smooth, chocolate-brown fusion crust of a freshly fallen space rock. For this reason the barren deserts of Earth, whether hot, stony plains on the periphery of the Sahara or bleak ice fields in Antarctica, have historically offered meteorite hunters the prospect of not just undisturbed meteorite falls but, perhaps more importantly, flat expanses of nothingness on which to find them.

My own quest for a more personal chunk of the solar system therefore brought me to Morocco, and I was fortunate to join a trip organized by the exploration team behind the Discovery Channel's *Meteorite Men*, who now offer professional tours. To the experienced hunters in my party, this landscape was better known as NWA—North West Africa—the three-letter code given to all meteorites found in the vast swath of territory covered by the rich prospecting grounds of the Sahara desert.

However, getting to Morocco would turn out to be only one half of the challenge. If I had learned where to search for meteorites, the question that now faced me on my first day in the field was how to tell the rich haul of meteorites that I felt sure was out there waiting for me from the intimidatingly large number of other rocks scattered on the seemingly endless stony plain stretching out in front of me. As a beginner, I found that the best time for meteorite hunting was early in the morning, with the light of a new day low at my back. In front of me a barren landscape stretched to the very foothills of the Atlas mountains. In the glancing morning light, the dark shadows cast by an uncounted multitude of rocks and pebbles receded and vanished. The plain colors of the rocks stood out, and dark rocks were particularly distinctive. To exploit this, I focused my attention anywhere up to thirty meters in front of me as I scanned the ground for rocks that looked like they just didn't belong there.

On any given day during our week of hunting, I would find myself a single speck in this endless landscape. Occasionally one of my companions would appear as a solitary, stick-like figure on the horizon as we quartered the ground between us. Searching for meteorites in this way feels Zen-like; one becomes insignificant compared to the scale of the landscape. A few times each hour I would come across a rock that didn't fit the local pattern—perhaps one that was a little smoother or darker than average—by just enough to catch a weary eye searching for a little hope. Most meteorites are a mixture of rock and iron—stony-irons in the parlance of hunters—and therefore my hardware-store magnet, professionally mounted on a cheap, wooden broom handle, was a trusted companion on my search. With a quick poke I could determine whether any meteorite-to-be at my feet was instead just another meteo-wrong looking to lead me astray. (Although nonmagnetic meteorites do exist, they are *very* rare. My back ached, and I was playing the odds.) If my promising rock was magnetic and equally smooth on its underside, I would venture to bend down and inspect it more closely. Those rocks that showed a smooth, varnish-like finish with faint surface cracks would definitely make it into my "finds" pocket for the journey back to camp and the expert opinions of my companions. I had been hopeful that I might have come across a rock that was chipped where a piece had broken off in flight to reveal the unmistakable signature of secondary fusion crust or even exposed interior.

I say hopeful because, after five days of searching, all I had to show for it was a few flecks of meteoritic iron and a newfound respect for the patience of meteorite hunters. Perhaps I should not have been surprised. Since scientists realized in the early 1990s that the Sahara might harbor up to ten thousand years' worth of meteorite falls, it has been picked over by everyone from goat herders, local traders, and Western tourists to scientific expeditions.

As I picked my own way across the desert from one interesting rock to another, I pictured the Sahara as seen from above and at a great height. Across the landscape stretched a human network of meteorite hunters, a pyramid of collectors, with humble goat herders and nomads forming its base. It is they who quarter the ground day after day, and season after season as they follow their flocks across the desert. For my brief experience in the desert, I could share their daily meandering but not replicate its scale across the years of their lives. Their finds would make their way to local dealers and thence to the larger towns of the region, where international meteorite traders would haggle back and forth with them in market scenes as old as humanity itself. Curious and cryptic rocks would be placed on scales and studied with jeweler's loupes. Chins would be scratched in contemplation of a potentially rare find, and hands would shake in completion of another deal. Should you wish to buy in bulk, you can pick up meteorites by the kilogram and play your stake in the "Morocco lottery," hoping that an unusual piece may have snuck its way into the batch of ordinary stony-iron meteorites that make up the majority of meteorites recovered in NWA.

I returned from Morocco with my own handful of meteorites—each a memento of a particular town market or bedouin tent—plus a few tiny fragments that bear witness to a bleak and dusty plain in Northwest Africa. Even the most common of meteorites represents an object of wonder when held in the hand. Each fusion crust is subtly different, an individual pattern of regmaglypts presenting a unique atmospheric fingerprint. The density is also different for each, a testament to the variety of asteroidal compositions from which they are drawn. Each meteorite that we find on Earth is a fragmentary clue—a solidified shard of memory from a near forgotten age of planetary assembly. Yet to learn more about meteorites and what they can teach us about life in our solar system we must cast our scientific net much wider—beyond the bounds of Earth itself. Scientists have identified distinct chemical and mineral patterns among meteorites, and these family relationships in turn point to the existence of parent bodies, perhaps still floating silently in the orbital realms between the planets. Visiting those parent bodies—asteroids and comets—has become a key objective for Earth's space agencies.

One might ask why we should go to the trouble of returning samples of asteroids to Earth when meteorites deliver pieces of them to us for free. The answer is that meteorites fall to Earth from a blind spot in our scientific vision—we don't know where in the solar system they originate. In rare cases the arc of a falling fireball is captured on dedicated cameras, and its trajectory can be extrapolated back in time to a possible location in the solar system. In other cases, such as the Howardite and Eucrite meteorites, their mineral properties provide a sufficiently close match to the observed surface properties of the asteroid Vesta that we can speculate that they are linked. Yet for the majority of meteorites studied on Earth we have little evidence to tell us where in the solar system they originated. We can make informed guesses, speculating based on the few clues we have. However, speculation rarely brings surprises. For that one needs to query nature directly, and in this case that means determining where meteorites come from. Sampling asteroids directly also offers solar system scientists the chance to learn about the detailed chemistry not just on the surface of asteroids but, crucially, within them as well. The chemistry of the carbonaceous chondrite meteorites in particular hints at the action of liquid water and at complex chemistry in the interiors of their asteroidal parents. These parent bodies were the building blocks of our solar system and, by extension, of Earth itself. If their chemistry turns out to be much more complex than originally thought, what might that mean for the chemical makeup of the young Earth and the prospects for life to emerge upon it?

It was this quest to sample material directly from likely meteoritic parent bodies that motivated two space agencies, NASA and its Japanese counterpart, JAXA, to initiate long-term programs to return samples from two of Earth's closer asteroids. Each agency had gained valuable experience in pathfinder missions as both NASA's *Stardust* and JAXA's *Hayabusa1* probes returned a few grains of material from comets and asteroids to Earth in 2006 and 2010, respectively. These endeavors paved the way for JAXA's *Hayabusa2* and NASA's *OSIRIS-REx* missions to attempt the more challenging task of sampling a few grams of material from the larger near-Earth asteroids Ryugu and Bennu.

It doesn't sound like much, but the acquisition of even a few grams of asteroidal material might allow researchers to determine the origins of some of the meteorite family groups. Any such links could then potentially unlock new secrets from the many hundreds of kilograms of meteoritic material secured in museum vaults across the Earth. It is this unique knowledge—priceless, one might say—that justifies the tremendous cost and effort involved in sampling solar system material directly from asteroid parent bodies.

The first of these probes, *Hayabusa2*, was launched in December 2014 toward the near-Earth asteroid Ryugu—a loose rubble pile of carbon and water-rich material just over a kilometer across. It is an Apollo asteroid, one of a group whose elliptical orbits overlap with that of Earth (yet fortunately pose little risk), and its composition and relative proximity to Earth meant that Ryugu offered the mission team at JAXA the right mix of scientific potential and a surmountable engineering challenge.

In Japanese folklore, Ryugu-Jo is the Dragon Palace, located deep within a mysterious underwater realm. In these traditional stories a fisherman, Urashima Taro, visits the palace, borne on the back of a turtle, and eventually returns to the surface world after the passage of many years bearing a cryptic sealed box. *Hayabusa2* completed its own rendezvous with Ryugu in 2018 and then spent the next few months executing a carefully choreographed orbital dance about the asteroid that allowed the probe to map its surface and select the optimal site at which to land and take samples.

As in any good fairy tale, the hero was aided in their quest by a pair of sidekicks. In this case they were two camera-equipped minilanders, *HIBOU* and *OWL*, each shaped like a flattened cylinder the diameter of a seven-inch cake tin. The two landers were designed to hop around the low-gravity surface of Ryugu using a self-contained

Ryugu from *Hayabusa2* The surface of the asteroid Ryugu imaged by the *Hayabusa2* probe. At this point the probe was just six kilometers from the asteroid, itself a little under one kilometer across. You can see why scientists commonly call such asteroids "rubble piles." The challenge for the mission scientists was to use images such as this to find a safe location to touch down and sample the surface. Credit: JAXA, University of Tokyo, Kochi University, Rikkyo University, Nagoya University, Chiba Institute of Technology, Meiji University, University of Aizu, AIST.

inertial mass and to take close-up images at each location. Over one month the two space hoppers sent 648 images of the surface of Ryugu back to Earth. They revealed a daunting prospect for the spacecraft team as the surface turned out to be littered with potentially mission-ending boulders and debris. This itself was a surprise, since it had been supposed that billions of years of micrometeorite strikes would have reduced the surface of the asteroid to a dusty regolith of small pebbles and grains. Months of analysis and replanning were required to select a site where *Hayabusa2* could safely descend to harvest a few grams of surface material.

Eventually, on February 12, 2019, the spacecraft began its slow and careful descent to the boulder-strewn surface of the asteroid. The surface gravity on *Ryugu* is utterly feeble, eighty thousand times weaker than on Earth. The escape velocity at the surface of the asteroid is one-third of a meter per second, less than walking pace, and *Hayabusa2* would need to approach the surface with exquisite precision to avoid immediately bouncing off again. At the precise moment when its extended sampling horn touched the surface, a specially crafted five-gram tantalum bullet was fired down the opening of the horn to strike the asteroidal regolith and kick a spray of material up into the collection chamber within the spacecraft.

It was a neat, if explosive, solution to the challenge of how to sample surface material with the minimum of moving parts. This is an important consideration for deep space missions, as each layer of complexity in the sampling plan represents a potential weak spot where a single failure could scupper the entire endeavor. With a few grams of surface material safely stowed on board, *Hayabusa2* could then contemplate its next sampling challenge: digging down deeper into the asteroid. The scientific motivation for doing so was clear, since material below the first few centimeters, tens of centimeters, or even a meter or more of the surface will have been shielded from the chemically altering effects of billions of years of ultraviolet solar radiation. The subsurface may well not be pristine, as plenty of interesting chemistry is expected to have occurred within the asteroidal interior, but it is important to study the traces of those chemical processes isolated from the effects of surface irradiation.

To achieve this goal *Hayabusa2* would have to dramatically up its excavation game—effectively using a bigger bang—without putting the spacecraft itself in jeopardy. Up until this point in its mission, *Hayabusa2* had used a relatively balletic suite of orbital maneuvers to dance around *Ryugu*, but now it was clearly time to deploy some fireworks. On April 5, 2019, *Hayabusa2* released a device known as the Small Carry-On Impactor at a precise location some 500 meters above the surface of Ryugu. This unassuming object consisted of 4.5 kilograms of explosive mounted behind a 2.5-kilogram convex hemisphere of copper. When detonated, the blast wave from the explosive charge would deform the copper shell into a fast-moving bullet-shaped projectile. Not only does copper offer the advantage that it is readily deformable, but it is not expected to be a constituent of the asteroid itself. The vaporized remnants of the copper projectile could therefore not be mistaken for material belonging to Ryugu. It was the perfect tool for the job, delivering a focused burst of kinetic energy to the surface of Ryugu with no moving parts required.

Hayabusa2 itself would not witness the moment of detonation. It left that job to a small fly-alone camera that would watch the expected impact point on Ryugu while the spacecraft itself took shelter on the far side of the asteroid. The impact of the copper projectile was satisfyingly effective—it excavated a ten-meter-wide crater and exposed subsurface material to a maximum depth of four meters. Further careful mapping of the crater was required to determine the precise spot to which *Hayabusa2* would descend to sample material from the crater floor, and on July 11, 2019, the spacecraft completed its second sampling maneuver at Ryugu.

With its visit to the Dragon Palace complete, *Hayabusa2* began its return journey to Earth in November 2019, and the spacecraft flew past Earth on a glancing trajectory some thirteen months later, in late 2020. True to the story of Urashima Taro, *Hayabusa2* returned many years after its departure to a world much changed in the intervening period. As the spacecraft silently sped past Earth, it released a small reentry probe containing the precious samples from Ryugu. *Hayabusa2* was able to look back and film the probe's descent as it streaked through Earth's atmosphere, becoming a shooting star in its own right before it gently touched down in the deserts of South Australia. The mysterious box of treasure, the scientific gift that *Hayabusa2* returned from Ryugu, was at that point safely in the hands of the JAXA recovery team and could complete its homecoming to a curation facility at Sagamihara in Japan.

Although it might not sound particularly impressive, the 5.4 grams of material *Hayabusa2* returned to Earth represented a veritable hoard of solar system knowledge. To put that quantity of material into perspective, in 2010 its predecessor, *Hayabusa1*, returned samples from the asteroid Itokawa that amounted to some 1,500 microscopic grains, totaling perhaps a little under a tenth of a gram. The initial analysis revealed that the samples returned from Ryugu bore a striking resemblance to the chemically primitive group of Ivuna-class carbonaceous chondrite meteorites— a result that was not surprising, as Ryugu had been chosen as the mission target with this expectation in mind. Yet, surprisingly, the Ryugu material was found to be relatively poor in glassy chondrites and calcium-aluminum inclusions compared to its meteoritic kin. Chondrites and metal inclusions are thought to result from the hot and violent conditions that prevailed in the inner regions of the forming solar system. Clearly, therefore, Ryugu had kept its distance from the sun during the early history of our solar system.

In a further twist, the structure of the Ryugu grains were found to be rich in a type of carbonate mineral that is deposited in interactions between liquid water and rock—a clue that points to a very particular sequence of events during its formation. Ryugu, or to be more specific the material from which Ryugu is derived, formed initially into a large asteroid, perhaps some tens of kilometers in diameter. Precise measurements of radioactive isotopes indicated that these events occurred exceptionally early in the history of the solar system, only some 2.6 million years after the formation of the oldest metal inclusions found in meteorites. Heat generated from the radioactive decay of the isotope aluminum-26 would have warmed the interior of the asteroid to somewhere between zero and 30°C. Liquid water was able to form as a result of these relatively clement conditions, and it precipitated a number of the mineral structures observed in the Ryugu samples today. However, that parent asteroid did not survive for long, being destroyed in a collision before some fraction reformed into the loose rubble pile that we know as Ryugu today. It is a wonderfully detailed story that adds new twists and turns to our understanding of the earliest phases of planet creation in our solar system. The initial results from *Hayabusa2* speak powerfully of the importance of seemingly fleeting environments in the early solar system and the role they played in establishing the chemistry and geology of the worlds that followed them.

And what of Urashima Taro himself? The stories relate

that Taro believed his journey had taken only a few days—but when he returned to his village he discovered that three hundred years had passed during his absence. Confused, he opened the cryptic treasure box given to him as a gift by a princess of the underwater realm only to be engulfed in a puff of white smoke. Taro emerged from the wisps of smoke aged and bent, and the secret of the box he carried with him was revealed to be no more nor less than his own old age.

The initial investigations of the Ryugu material returned by *Hayabusa2* have provided ample demonstration of why science should go to the trouble of returning samples to laboratories on Earth. It is also exciting to note that the success of this mission marks only the first chapter in the story of asteroid sample return that is playing out in real time around us. In September 2023 NASA's *OSIRIS-REx* mission returned its own samples to Earth from the carbon-rich asteroid Bennu, and the coming decade will host a golden age of asteroidal discoveries. It has been and will be a painstaking process. Our present knowledge of the bodies of the solar system is patchy at best—a planet here, a moon there, with a scattering of rocky fragments in between. Yet piece by piece, asteroid by asteroid, then later perhaps moon by moon and planet by planet, scientists will slowly but surely stitch the physical history of our solar system together into a coherent whole. Their aim will be to understand how our system of planets—and ultimately ourselves upon them—came to be.

The tiny, yet precious, samples of asteroids returned by *OSIRIS-REx* and *Hayabusa2*, together with the bounty of meteorites that fall to Earth, provide the crucial primary evidence used to reconstruct the earliest history of our solar system. There are many fascinating stories that we could tell, but we remain focused on those tales relevant to our Earth-bound perspective on this pale blue data point. To that end, what does the evidence tell us about the possibility for life to emerge in our solar system, either on Earth or beyond it?

Two questions stand out that must be answered in order to understand how life came to develop on Earth: Where did the water that makes up our oceans come from, and what was the chemical process that synthesized the carbon-bearing molecules that formed the precursors of life? To answer these questions we have to go back to the earliest phases of planet formation in our young solar system. The classic view is that our sun formed at the center of a collapsing cloud of gas and dust. As this pre-solar nebula fell under the force of its own gravity, the almost imperceptible rotation of the cloud was magnified to form a spinning disk of material in orbit about our nascent sun. We observe such protoplanetary disks around young stars today. The bright Southern Hemisphere star Beta Pictoris hosts one of the examples closer to Earth, and we will discover other, astonishing examples later in this chapter when we travel to the ALMA Observatory in northern Chile. The inner regions of these disks are heated by their young stars, with the temperature of the gas and dust declining steadily toward the outskirts. Water vapor is a common gaseous ingredient in protoplanetary disks, along with a host of other basic molecules such as carbon dioxide, ammonia (NH_3), and hydrogen cyanide (HCN).

If an imaginary astronaut traveled through the outskirts of one of these disks, which are at this stage still dominated by a mixture of gases and microscopic dust grains, they would pass through a zone known as the frost line, the point at which the declining temperature of the disk allows water vapor to condense directly and form ice on the surface of dust grains. In our own solar system the frost line is located between the orbits of Mars and Jupiter, approximately midway through the asteroid belt. It was long assumed that planets forming interior to the frost line would be desiccated and rocky, depleted of volatile gases such as water vapor compared to planets formed beyond the frost line. This line of thinking led to the preconception that Earth was born dry—a depleted world devoid of water. If that assumption was correct, however, then where did Earth get its water from? Many today still look toward carbonaceous chondrite meteorites as the source of Earth's water and other volatile compounds—delivered to the young Earth during a protracted period of asteroidal impacts. The key piece of evidence that supports this view is that both the Earth and this primitive class of meteorites possess the same mix of heavy isotopes in their reservoirs of water. Among every ten thousand atoms of hydrogen that exist in water molecules on Earth, there will be found one or two atoms of the isotope deuterium. Whereas regular hydrogen contains one proton in its atomic nucleus, deuterium carries within it a neutron in addition to the single proton that is normally present. This ratio between hydrogen and deuterium in Earth's oceans does indeed match that found for water in carbonaceous chondrite meteorites, and at first glance this suggests a physical link between the two. However, other, much rarer, chemical elements in carbonaceous chondrite meteorites, such as xenon and osmium, do not match the isotopic ratios of those elements when found on Earth. Although this finding does not disprove the existence of a link between the young Earth and the parent bodies of these meteorites, it does mean that the chemical story may not be as simple as we first thought, and we may need, quite literally, to dig deeper to discover more.

At this point we turn our attention away from meteorites for a moment to ask instead what clues we have regarding the conditions that existed on the young Earth. Was it indeed born dry, as some think, or does the evidence point to a different beginning for our planet? As ever in our investigation of the earliest history of both Earth and the solar system, we will discover that the evidence we seek is written in rocks. The chunks of meteorites that we can hold in our hand tell us that the solar system was formed just over 4.5 billion years ago. Yet the oldest intact rocks on the surface of Earth are a fraction over 4 billion years old and are found in a small outcrop on the Acasta River, about 300 kilometers north of Yellowknife in Canada's Northern Territory. Careful prospecting may one day reveal older rocks, but for the moment these are the oldest record of Earth's surface available to geologists. The older rocks, stretching back to Earth's molten genesis, are presumed either to be buried beneath the surface or to have been drawn down to Earth's mantle where they have melted and been reformed as new rock. But that geological quirk of circumstance leaves scientists with an embarrassing half-billion-year gap in the story of Earth.

That story is slowly being filled in, but not by any rock that you or I can hold in the palm of a hand. Instead, geologists have learned to decipher the clues held within ancient and tiny mineral crystals of zircon—great survivors from the dark ages of Earth's prehistory. Zircon itself is an exceptionally tough mineral crystal composed of zircon, silicon, and oxygen and is described by the chemical formula $ZrSiO_4$. The most ancient of these crystals, whose ages extend to an astonishing 4.4 billion years old, have survived episodes of erosion and melting that completely erased the original rocks that once held them. Each is typically the size of a grain of sand, perhaps a tenth of a millimeter across, and represents a speck of geological memory that recalls a primeval world. In the same manner that an archaeologist can reconstruct the lives of past humans from fragments of pottery and the detritus of their homes, so can an isotope geochemist use the ratios of radioactive isotopes contained in these zircons to reconstruct the history of Earth in the Hadean eon that describes our planet prior to 4 billion years ago.

The story of oxygen isotopes within the oldest of these zircon crystals tells us about the earliest history of the surface of Earth. Presented in a 2001 *Nature* article titled "Evidence from Detrital Zircons for the Existence of Continental Crust and Oceans on the Earth 4.4 Gyr Ago," this revolutionary analysis determined that a zircon crystal containing these oxygen atoms had itself formed from melted crustal rocks that existed in contact with liquid water. However, the image of minerals precipitating from a water-rich environment 4.4 billion years ago challenged the long-held view that during the Hadean eon Earth was a hellish, molten world. Oxygen isotopes in ancient detrital zircons instead tell a story of an early Earth with a cool, solid crust and significant quantities of liquid water. Mark Harrison, an expert in early Earth isotopic geochemistry and an author on a companion paper to the 2001 *Nature* article, has enjoyed pointing out the irony of the fact that the original Greek conception of Hades was of a dark, cold, watery realm. Contrary to long-standing thought, the Hadean Earth may have been "born wet" and experienced a long history of liquid water on its surface. Using the modern astrobiological definition of the word, Earth was habitable during this earliest of epochs. Perhaps not continuously, though, as the solar system remained a turbulent and violent environment. Yet fleeting conditions on the early Earth may have allowed the great chemistry experiment that we call life to have got underway much earlier than previously thought.

So where does this discovery leave meteorites and the dry Earth hypothesis that I introduced earlier? Evidence from ancient zircons points to the possibility that the early Earth possessed significant amounts of liquid water and retained it long enough to leave clues to its presence in the geochemistry of its surface rocks. When new evidence challenges existing ideas, one has to return to the assumptions those ideas are based on. In this case it means asking whether the assembling Earth might have retained a dense, steamy atmosphere that condensed as the young planet cooled. Did Earth's deliveries of water derive from water-bearing minerals accreted during its initial formation or perhaps from the same water-rich asteroids that gave rise to the carbonaceous chondrite meteorites that we observe today? Was the water they transported shielded from high temperatures interior to the frost line, perhaps within floating grains, pebbles, and rubble piles? It is a wonderful example of the power of primary evidence to challenge speculation, because whatever happened to the water on Earth, detrital zircons tell us it happened early. Does this uncertainty over the role meteorites played in seeding Earth's oceans diminish their importance to astrobiology? Should we just toss them to one side? Certainly not. These meteorites probably still represent the material that contributed to the accretion of Earth and its reserves of water—the current question facing scientists is exactly *when* that accretion occurred.

Because of their primitive nature, carbonaceous chondrite meteorites have other stories to tell us regarding the very earliest conditions in the solar nebula, in particular the story of nature's ability to perform complex chemistry in the depths of space. The evidence for an early ocean on Earth may also have important implications for what other volatile compounds were present on our young planet at the same time. For if water could be retained during Earth's formation, then so might other compounds have been preserved, a chemical inheritance that may have had important implications for the journey to life on Earth.

A typical carbonaceous chondrite meteorite contains about 2 percent of its mass in the form of organic carbon compounds. Simple molecules such as formaldehyde (CH_2O) are more common than more complex forms such as amino acids. Yet an incredible eighty-six varieties of amino acids have been detected in the Murchison meteorite that fell to Earth in 1969. Furthermore, a great deal of the carbon in meteorites is in the form of amorphous microscopic agglomerations that are not identified as any particular molecule. Where did this astonishing variety of stuff come from?

The meteorites we observe today, however, are essentially inert fragments, and the chemical and mineral evidence they preserve instead implies the existence of much more active conditions within the ancient asteroids from which they are derived. The samples returned by *Hyabusa2* and *OSIRIS-REx* have taken us one important step further in the story of these meteorites: They have collected pristine material from within their likely parent bodies. The early results from the analysis of these samples point to liquid water seeping through the warm interior of these fragile, rocky worlds to form a salty broth rich in organic chemicals. It is deeply compelling to consider that the current results from these precious samples represent only the very first of many laboratory investigations that surely presage more wonderful surprises to come. Yet these same results also allow us as astrobiologists to imagine the ancient planetoids, of which worlds such as Ryugu and Bennu are merely the present-day fragments, as primordial chemical factories in which nature was allowed to perform an intriguing and provocative variety of chemical experiments.

If life itself can be viewed as a chemical system that has passed a critical threshold of complexity, then carbonaceous chondrite meteorites demonstrate that our solar system started out on that path to chemical complexity very early indeed. They tell us that life on Earth perhaps did not have to start exactly from zero, that meteorites might have seeded Earth with complex organic material that then reacted further to create chemical cycles of sufficient complexity for us to label them as life. Of course it is important to consider the alternative view, that Earth might well have formed with a relative dearth of such complex organic material and that it later synthesized these chemicals naturally via atmospheric and oceanic processes on the surface of its young and water-rich crust. At present we lack the evidence to favor one explanation over the other.

What we can say for certain is that meteorites and their asteroid parent bodies contain evidence of complex chemical cycles, and for that reason alone they remain of particular interest to astrobiologists. We can turn back the clock and imagine an atom of carbon trapped within an asteroid tumbling through our young solar system. The warm, aqueous interior of the asteroid serves to jostle our carbon atom through a maze of chemical reactions. The paths it takes will be determined by the different types of bonds that it can form with its neighbors and the ability of those bonds to resist being broken apart again in the bustling molecular crowd. Some paths are cyclic and loop back upon themselves in repeating patterns of formation and dissolution. Others lead to rare havens where a particular molecule might achieve a very stable form and resist change.

The chemical reactions that are believed to have occurred in the parent asteroids of carbonaceous chondrite meteorites echo those of the famous Miller–Urey experiment. In 1952 Harold Urey, himself a Nobel laureate in chemistry, directed his then graduate student Stanley Miller to see what would happen if a mix of chemical ingredients representing the atmosphere and oceans of the early Earth were let loose in a labyrinth of sealed beakers and flasks to explore the multitude of ways in which they could form stable compounds. Life was not the preordained goal of a chemical system such as we believe existed either within the asteroids of the early solar system or on the surface of the Hadean Earth. However, many scientists believe that what we call life emerged as a by-product from a process similar to the one I just described—when a bubbling broth of linked, cyclic chemical reactions passed an unknown but critical threshold of complexity en route to forming carbon-bearing molecules of greater stability. Although it is unlikely that this threshold was passed in asteroids, the first steps of a long and winding chemical journey may well have been taken there. If these volatile organic molecules were preserved on Earth, then further chemical reactions may have occurred in the bubbling aqueous margins where water met hot rock on the young planet. Viewed in this way, neither the chemistry of asteroids alone nor that of the early Earth should be considered as telling the whole story of the chemical route to life on Earth. What is important, however, is that meteorites provide scientists with a material record of that chemistry. On Earth we so far have no such record of those processes—although the study of carbon-rich inclusions in detrital zircons may one day provide tantalizing glimpses of carbon chemistry on the Hadean Earth.

Such is the story of our young solar system and the chemical prehistory that preceded the emergence of life on Earth. Prospecting for clues among the asteroids has brought that story to the cusp of a spectacular new age of discovery. Yet the next step in our journey together will be equally astonishing. For with one bound we will leave behind the asteroidal relics that reveal, if only dimly at times, the earliest history of our solar system and land firmly and squarely before remarkable protoplanetary systems forming about new young suns. Staying true to the traditions of Urashima Taro, this remarkable journey will take only as much time as is needed for an array of radio telescopes to trace their arcs across the sky. And for me the trail will lead back to the mountains of Chile to visit a quite unique observatory.

My first encounter with the Llano de Chajnantor was over twenty years ago, before it became the site of a major international observatory. I had visited the area as part of a team of astronomers investigating the nearby mountains for suitable sites to build a future telescope. The promise of Chajnantor had brought us to the apex of the central Andes, the summit of a great pyramid of land that lies close to the borders of Chile, Argentina, and Bolivia. It is an austere location, a barren, stony plateau set at an altitude of 5,000 meters above sea level. A ring of 6,000-meter mountains circles the vast plain, yet, when viewed from the expanse of Chajnantor, they appear as low, bare hills in the vertically compressed geography. An immense blue sky elevates the scene, seeming to lift the observer above the ground, to float, detached, in the clear, intangible air. In the absolute quiet, broken only by the sighing of the wind, one feels closer to the all-encompassing sky than to any place on Earth. By day we scrambled up rocky slopes to install remote weather stations on the barren peaks surrounding the plateau. Yet at night we found ourselves lifted to the level of the stars as we used a suite of portable telescopes to measure the almost perfect clarity of the atmosphere above us. From our vantage point, one seemed to look not up at those stars, distant and unattainable, but out and across to them, flickering bonfires of luminosity on neighboring cosmic peaks.

Chajnantor today is the location of ALMA, the Atacama Large Millimeter Array. The daytime sky above it remains just as I remember it, a vault of the most perfect porcelain, without flaw or blemish. Spread beneath it is an array of sixty-six radio telescopes deployed across the length and breadth of the plateau. If one sat on a nearby peak to take in the view, the white antennae would appear to dot the stony plain below, the farthest some sixteen kilometers away. A wandering mind might recast the scene as a modern version of some megalithic observatory—a Stonehenge or Carnac. From that distant perspective each antenna appears as a tiny sentinel, yet up close each structure is a 115-ton monolithic giant. Most of the antennae feature a steerable parabolic dish twelve meters in diameter, and in the past each of these stand-alone telescopes would have constituted a radio observatory in its own right. However, each imposing ALMA antenna can be moved independently about the plateau by a pair of custom, twenty-eight-wheel heavy transporters, and the entire observatory is wired to operate as a single extended telescope. It is in this way, operating as an interferometric network, that the true power of ALMA can be appreciated.

The power of interferometry lies not in the combined collecting area of the telescopes but in the subtly different perspective that each antenna provides of objects in the night sky. Just as the ripples produced by a stone tossed into a pool arrive at the surrounding shore at different times, so does the light from a distant astronomical source arrive at an array of telescopes with a slight delay. By compensating for this delay, astronomers can create an exceptionally sharp image of the target, far in excess of what can be achieved with a single antenna. Using this technique, an interferometric array of telescopes can reproduce the resolving power of a much larger telescope without the need to construct a single telescopic giant. In this sense ALMA offers a technological counterpoint to the Extremely Large Telescope we visited earlier, with each facility providing complementary and utterly unique views of the universe.

So far in our story of planet hunting we have discussed how the radial velocity technique can reveal the reflex motion of hidden worlds. In addition, adaptive optics imaging can directly, though barely, reveal planets in orbit about their parent stars. Compared to these techniques, ALMA offers astronomers the truly novel ability to observe the birth of worlds—nascent planetary systems forming in warm, dusty disks that orbit capricious young stars. The star system HL Tau was the first such system to be revealed by ALMA, in 2014, and the image it produced remains one of penetrating scientific impact.

HL Tau is a young star in a complex star-forming region in the constellation of Taurus. It is one of a class of stars that are in their teenage years, on the cusp of adulthood, when nuclear fusion commences in the stellar core and the star joins what astronomers refer to as the main sequence. The so-called T Tauri stars are fickle and variable as they are in the process of contracting and generating the vast quantities of heat required to ignite their nuclear fires. HL Tau is perhaps only 100,000 years old, certainly no more than one million. From as far back as the 1980s it was known to be surrounded by a rotating disk of gas and dust, yet the exceptional angular resolution and sensitivity provided by ALMA revealed something truly extraordinary. At greatly magnified scale, the disk of HL Tau was revealed to be rather like an inclined vinyl record, complete with concentric circular bands of bright emission interspersed with dark gaps. The whole image appeared not unlike the rings of Saturn, yet these features represent not the motion of moonlets and icy fragments in orbit about a planet but an entire protoplanetary disk in motion about a young sun perhaps no more than 100,000 years old.

The image was immediately astonishing to both experts and the public alike, and, rather like the earlier image of the pale blue dot, it spoke of the unique power of a single astronomical image to redefine our view of our place in the cosmos. The first startling revelation was the fact that HL Tau is so young, yet its disk seems so ordered, implying that protoplanetary disks form quickly and vigorously, in step with the forming star and not as a sedate afterthought. The ordered sequence of bright and dark bands hinted at the presence of forming planets in the process of gravitationally scouring their orbital lanes about the star. One should note that there is as yet no direct evidence of planets, and it might be that we are observing waves of pressure in the disk, caused by the interplay between gas and

HL Tau A single astronomical image of deep emotive power. This image of the gas and dust disk in orbit about the young star HL Tau was taken by the radio antennae of the ALMA Observatory in northern Chile. The concentric rings of emission span a diameter of some twenty-five light hours, about two and a half times larger than the orbit of Pluto in our own solar system. Astronomers speculate that the gaps in these rings may mark the orbits of forming planets. The glassy chondrules in my slice of the Allende meteorite were likely formed when our own solar system looked like this one. Credit: ALMA (ESO/NAOJ/NRAO), NSF.

dusty pebbles. Even so, these features are thought to mark the sites where planets may well form in the future, as the bunching up of material in these bright orbital bands is thought to accelerate the process of collision and growth. In this sense, the image of HL Tau is impressive both for what it reveals directly and for what it only yet hints at.

It is also important to consider the wavelength at which these observations are performed and how they illuminate the physics at work in the HL Tau disk. The image of HL Tau was created using photons at a wavelength of 1.3 mm, and light of this nature is produced by warm dust and pebbles in a largely transparent disk. However, I should clarify that when astronomers use the word "warm" they mean a few tens of degrees above absolute zero, and "pebbles" refer to lumps of material from a few millimeters across to rocks that could fit in the palm of your hand. If you thought that sounds very much like the conditions we believe existed in our solar system around the time of the very earliest phases of planet formation, then you would be right. Put simply, we are potentially witnessing, in a very direct manner, the very first phases of planet formation, akin to our solar system when it formed the calcium-aluminum inclusions and chondrules that we find in carbonaceous chondrite meteorites.

As more and more protoplanetary disks have been observed using ALMA, they have revealed a breathtaking variety of disk structures. Warps, cavities, gaps, arcs, and spirals have all been observed, features that taken together represent a kaleidoscope of planet-forming possibilities. The tale is one of incredible variety and tremendous surprises. In one sense we should be surprised that we are surprised. Our first views of exoplanets, now almost thirty years old, were characterized by a similar abundance of variety. Perhaps the most striking aspect of our current view of exoplanets is how uncommon are systems that resemble our own solar system, even given the technical limitations that affect our ability to detect all types of planets about all types of stars. ALMA has proved to be a discovery machine of the first order. It has taken us to realms that previously were either blanks on the map or at best only sketched in outline. The nascent worlds it has revealed are spectacular in ways that we could only have guessed at more than twenty years ago when I first sat on a mountainside and gazed over the Chajnantor plateau.

Taken together, the ALMA observations of protoplanetary disks compiled over the past ten years allow us to tell a detailed story of our own origins. In relative terms these disks are short-lived, lasting only some 10 to 15 million years following the formation of their parent star. After that they apparently dissipate—to be replaced by young planets and a few glints of debris. We can place the end of the age of disks in context by using precise dates from our own solar system. The oldest recorded event in our solar system, the formation of calcium-aluminum inclusions, occurred 4.568 billion years ago. Yet the maximum ages of disks observed by ALMA imply that planet formation may have been largely complete in our own planetary system after only 10 million years, or 4.558 billion years ago—the challenge here is not to get lost in the decimal places of these precise numbers! Perhaps 100 million years later, 4.458 billion years ago, a comparison of radioactive isotopes of elements such as lutetium and hafnium found in meteorites and in Earth's crust reveals that our molten planet had differentiated into a metallic core and a rocky mantle. As we have seen, detrital zircons in Earth's crust have revealed that as much as 4.4 billion years ago the surface of the Earth was home to stable reservoirs of liquid water. At its shortest, this interval of time is 168 million years after the very first phases of planet formation itself. That interval may appear to be very short to you, and perhaps it is astonishingly short given our previously speculative views of what conditions were like on the surface of the Hadean Earth.

Yet it is supported by a compelling and diverse assembly of primary evidence.

What does the story told above mean for life on that early Earth? More than anything else, it provides the gift of time—the young Earth may have offered a habitable environment far longer than we have previously assumed. The interesting question, of course, is what might life have been doing in that dark, Hadean age. It is intriguing to speculate whether life may have arisen multiple times on that young Earth, each a hopeful roll of the chemical dice, only to see those hopes dashed in a local or even planetary-scale cataclysm. The life that we represent today, which has existed since our earliest accepted fossil and geochemical evidence, may represent the fortunate lone survivor of an as yet unglimpsed eon of biological experimentation. From this literal primordial soup of possibilities we may have emerged as, perhaps not the sole victor, as we may represent a patchwork of recipes for early life, but certainly as the lone biochemical memory of that early time.

The meteorites that we have encountered in this chapter are the great survivors of our solar system—each a physical fragment of cosmic history that comes to us not quite from a forgotten age but certainly from one only dimly remembered. In one sense this is a strange statement to make. We know that all matter in the universe was created 13.8 billion years ago in the Big Bang. One could therefore claim that every object around us is the same age and that no one piece could claim the title of being the oldest chunk of the universe. Yet there is a more subtle point to be made here. All matter is subject to change—stars flare into existence within swirling clouds of gas. Planets grow around these stars—some to be destroyed in violent collisions while others cool and coalesce into potential habitats for life. Supernovas explode and cast nuclear ash into the void of space from which new stars, perhaps our own sun among them, will one day form.

So when I say that a meteorite is a tangible relic from ages past, I mean that it is a solid object whose form has remained unchanged over periods of time that, while perhaps not as long as the age of the universe, may extend to the age of our solar system and even slightly beyond. Each is a visitor to our home planet—a guest come to sit around our earthly fires and, in the tradition of Homer, repay the gift of hospitality by sharing with us stories of their travels. The stories told by samples of the asteroids Ryugu and Bennu have been and will continue to be rich and intricate. Their woven tales will allow us to re-create the earliest fabric of our solar system. Other visitors, among them the meteorites and fragments of interstellar dust, also appear in equally Homeric fashion, yet they more closely resemble ghostly shades on the edge of the firelight, revealing their hidden stories only partially, in riddles and whispers, as the gods dictate.

That we as astrobiologists can now glimpse other solar systems in the act of formation provides us with a distant but direct perspective from which to interpret the fragmentary clues we have from our own solar system. When interpreted together, they may begin to reveal the answers we have sought about how Earth emerged as a life-bearing planet. Our task, much like Homer's almost three millennia ago, is therefore to draw together all of those stories into one magnificent epic that will tell the tale of our origins in time and space.

6

SO LONG AND THANKS FOR ALL THE FISH

A DOLPHIN-LED GUIDE TO ALIEN COMMUNICATION

> Man has always assumed that he is more intelligent than dolphins because he has achieved so much—the wheel, New York, wars and so on—while all the dolphins had ever done was muck about in the water having a good time. But, conversely, the dolphins had always believed that they were far more intelligent than man—for precisely the same reasons.
>
> DOUGLAS ADAMS, *The Hitchhiker's Guide to the Galaxy*

Douglas Adams would not, I am sure, have been surprised at the role that dolphins have played in the development of modern astrobiology. In the spirit of the pale blue data point, dolphins present us with a compelling terrestrial opportunity to confront our ideas of what constitutes intelligence and, perhaps more challenging, just how might we recognize it. They provide us with one path to tread as we explore the later stages of the Drake equation. For our journey together has brought us to the very edge of the astrobiological map.

Up until this point we have considered the question of what constitutes life at the fundamental level of biochemistry and how, as scientists, we might recognize it. Given that the search for life is the ultimate, one might say only, goal of astrobiology, focusing on that preeminent question makes good sense. Yet the emergence of life represents only one term in the Drake equation, and it is not even the final one. Later terms represent the emergence of intelligence and the ability to communicate.

The Drake equation has been interpreted in many different ways since Frank Drake first wrote it down more than sixty years ago. The most common viewpoint is that it is a prism that refracts the continuous story of life in the universe onto discrete mathematical terms—akin to a series of checkpoints, each to be marked off on the cosmic journey to complex life. Another way to view the equation is as a parable of our own existence—the story so far, if you will. The invention of powerful radio transmitters in the second half of the twentieth century awoke in the minds of a number of scientists (and authors, for that matter) the latent idea of communicating with alien civilizations. In attempting to quantify the unknown challenges involved in that search, the Drake equation inadvertently tells the story of how we got to the point where we became capable of searching.

The search for intelligent life, to discover the supposed advanced civilizations of science fiction, appears audacious when compared to the more methodical steps we have so far taken toward the discovery of basic life. Yet SETI, the search for extraterrestrial intelligence, makes up the very DNA of astrobiology and has been present since the very beginning. One could even make the argument that the modern science of astrobiology came into being at a small academic conference at the Green Bank Radio Observatory in the autumn of 1961. Not only did this three-day meeting witness the advent of the Drake equation as the defining idea of astrobiology, but the attendees themselves—all ten of them—represented, from the moment of birth of this new science, its interdisciplinary character.

There was clearly something in the air during the late 1950s, or perhaps we might better say that it was instead in the ether, the imaginary medium through which electromagnetic waves were once thought to travel. In 1959 Guiseppe Cocconi and Philip Morrison, a pair of groundbreaking nuclear and particle physicists with little or no experience of radio astronomy, had written a seemingly bizarre academic paper that articulated how a search for interstellar communication might be performed using the radio technology of the day.

Earlier that same year Frank Drake, then a recently graduated astronomer, had proposed using the twenty-six-meter radio telescope at Green Bank Observatory to search two nearby sunlike stars, Tau Ceti and Epsilon Eridani, for alien radio signals using the exact method that would later be outlined by Cocconi and Morrison. Observations were due to start in 1960, and both Frank and his boss at Green Bank, director Otto Struve, had originally decided to avoid publicizing the project for fear of ridicule. However, upon seeing Cocconi and Morrison's paper in the journal *Nature*, they decided that their cover had been blown, and Project Ozma was born.

At the core of their common method lay the simple idea that if you were to pick one ubiquitous property of matter that might serve as a beacon for communication, one that all intelligent species would know of, no matter what their location within the Milky Way, then you might well have chosen the hydrogen 21-centimeter line. This feeble yet pervasive atomic glow is emitted by low-density clouds of hydrogen gas that fill the space between the stars within our galaxy. The weak flash of energy released as hydrogen's sole electron performs an acrobatic flip from one quantum state to another is manifest as a single photon of wavelength equal to twenty-one centimeters or, if you prefer, of frequency equal to 1.42 gigahertz. It is a rare transition, typically observed only in very low-density clouds of gas, and was first observed by Earth-based astronomers in 1951. Yet hydrogen can be found everywhere in our galaxy. We humans had discovered it using the most primitive of technology, in this case a radio receiver hung precariously outside a Harvard University laboratory window in a manner akin to a cosmic air conditioning unit. The 21-centimeter emission that we observe throughout our galaxy is not manmade, nor even alien-made—it arises from the physics occurring in the interstellar hydrogen gas within the Milky Way. Yet—assuming of course that the aliens think as we do—it might provide just the kind of electromagnetic beacon about which we could broadcast signals of our own.

Once detected, hydrogen 21-cm emission proved to be a powerful cosmic idea, a spark that lit a flame in the minds of not just Cocconi, Morrison, and Drake but others as well. The culmination of Cocconi and Morrison's thought experiment was to imagine whether that presumably well-recognized spot in the electromagnetic spectrum occupied by hydrogen emission could serve as a cosmic communication channel. They calculated that an alien radio telescope, broadcasting close to 21 centimeters, could be detected by our own facilities if it was located within the nearest star systems to Earth. The fact that Frank Drake was, at the very same time, preparing to begin his own search along much the same lines as would be drawn by Cocconi and Morrison speaks of a moment of pure synchronicity, the physical realization of an idea whose time had clearly come.

Others would share this idea and would be inspired by it. Among them were Joshua Lederberg, who sat on the Space Science Board of the National Academy of Sciences—an influential body that advised the young NASA. It was he who floated the idea of a conference to investigate the idea of the search for extraterrestrial intelligence, or SETI, as it would become known, in an effort to harness these bold and imaginative new ideas to the future exploration of the solar system. The Green Bank conference was the result of this early flash of scientific opportunity. Peter Pearman, Lederberg's associate on the Space Science board, and Frank Drake were tasked with drumming up interest in the meeting. Ten scientists would eventually heed the call, and their ideas would reverberate through astrobiology for decades to come.

In addition to Pearman, Drake, and Struve, the conference was attended by Philip Morrison and Su-Shu Huang, a former graduate student of Otto Struve's and an early expert in the nascent search for exoplanets. A young Carl Sagan, freshly graduated from the University of Chicago with a PhD in planetary science, was a driven and eager confidant of both Frank Drake and Joshua Lederburg and one for whom the Green Bank meeting could not have occurred at a better time. Also present were Dana Atchley and Barney Oliver, two technology experts who had helped Frank Drake make Project Ozma a reality.

Up until this point there was a strong astronomical flavor among the attendees, although I do find it interesting to consider what scientific fields were *not* represented at

Green Bank in 1961. There were no geologists or oceanographers, for example. In 1961 these fields remained poised on the brink of their own great revolutionary decade. The science of planetary geology and the perspective of Earth as just one planetary possibility among many had not yet been born.

However, the presence of two more attendees at Green Bank testified to the true interdisciplinary nature of the scientific field they were in the process of creating. Melvin Calvin was a chemist at the University of California, Berkeley, who in 1950, with Andrew Benson and James Bassham, had discovered the metabolic carbon cycle that lies at the heart of photosynthesis. Calvin and his associates would be named as winners of the 1961 Nobel Prize for chemistry during the Green Bank meeting. From the beginning, therefore, Calvin's presence at the meeting marked a strong connection between astrobiology and the search for basic life using biochemical insights.

The final attendee was John Lilly, who in 1961 was already a deeply influential polymath. His early medical career had focused on neurophysiology, a field of study for which he invented the isolation tank as a research tool. By the mid-1950s, however, he had begun to study the neurophysiology of bottlenose dolphins and their complex vocalizations. His research led in 1961 to the publication of his quite astonishing book *Man and Dolphin*, in which he speculated on the intelligence that might underlie dolphins' large, complex brains and on the implications this would have for our own relationship with them. Some have commented that, if the other attendees at the Green Bank conference were famous, then Lilly was the only one who was *infamous*. Although Lilly's research frequently skirted the fringes of the scientific pale, it was only later that he would embrace more fully the scientific counterculture of the 1960s, when he would become consumed by his fascination with the neurological potential of powerful hallucinogens. However, in 1961 he remained a deeply original thinker whose ideas clearly resonated with those of the other attendees at Green Bank.

At the very beginning of the meeting, Frank Drake wrote the individual terms of his equation on a blackboard. He expressed them in the convenient shorthand of mathematics so that they might serve as an agenda for the discussions that followed. Interestingly, apart from the later personal recollections of some of the participants, the Drake equation remains the only written record of this seminal meeting. To be fair, accounts of the meeting make it sound like a three-day scientific stream of consciousness, an environment in which ideas would fuse, break apart, and re-form as the participants flowed from the discussion room to meals and back again—not to mention enjoying the champagne-fueled celebration of Calvin's Nobel Prize. But there was one other physical record of the Green Bank meeting. The attendees had whimsically dubbed their gathering the Order of the Dolphin in honor of John Lilly's work, and after the meeting Melvin Calvin commissioned a set of dolphin-shaped lapel pins to present to his colleagues and formalize the scientific companionship the meeting had fostered.

Frank Drake's Project Ozma was the first SETI search of the modern era. Since that time many technological innovations have been applied to improve the sensitivity of radio searches and multiply the number of radio frequencies that can be simultaneously monitored. Whereas Frank Drake listened for signals from just two sunlike stars,

Epsilon Eridani and Tau Ceti, over a period of four months, a present-day radio SETI search monitors some several hundreds of stars, over a much wider range of frequencies, in the same period.

Drake used the twenty-six-meter radio telescope at Green Bank to search for anomalous signals emanating from these two stars. While pointing the telescope at each star he used a radio receiver sensitive to a band of frequencies just 100 hertz in width to map out a total of 400 kilohertz (kHz) of radio emission centered on the magic 1.42 gigahertz (GHz) hydrogen emission line. The radio energy received from each star was recorded as a line drawn on a rolling drum of graph paper and also played over the control room loudspeaker. The process he followed was not unlike using an old-school radio tuner to search for your favorite FM radio station. The FM band for commercial broadcasts covers frequencies from 88 to 106 megahertz (MHz), yet any single radio station is broadcast in a narrow 0.2 MHz band within this range. To those of us used to the analog car radios of yesteryear, finding our favorite station would involve gently scanning the tuning dial close to the target frequency to find the right spot, followed by some careful fine-tuning to get the best signal. Frank Drake was using the Green Bank telescope in a similar manner in his quest to tune in to alien broadcasts—except that he would have had to retune his receiver four thousand times for each star in order to cover a 400 kHz interval with his 100 Hz receiver.

Drake scanned the radio receiver back and forth as he scythed through the fields of static received from each of the two stars. He later wrote that this was a dispiriting business; the rush of searching for alien signals would quickly dissipate and be replaced by a fug of boredom. Drake went so far as to counsel later SETI searchers not to give up their day jobs in other fields of astronomy—not, as you might think, in order to avoid ridicule, but instead to find intellectual relief from the doldrums of hearing nothing but static for weeks on end.

We can compare Frank Drake's efforts with Project Ozma to a more modern SETI search, one undertaken as part of the ten-year Breakthrough Listen project and published in 2017 in the *Astrophysical Journal*. The research team used a newer telescope at Green Bank, one with a diameter of 100 meters, to observe a total of 692 stars. The real difference between the two projects, however, was the use of a modern, multichannel radio receiver that could simultaneously monitor the frequency interval from 1.1 to 1.9 GHz in frequency slices as narrow as 3 Hz. Expressed in terms of what we might call discovery power—the number of stars observed multiplied by the width of the searched frequency interval—the 2017 Breakthrough Listen article described a search that was some 350,000 more effective than Project Ozma. In addition, this particular report described only one part of a much more ambitious project to eventually observe all of these stars over the total frequency interval from 1 to 12 GHz.

Despite these two studies, and all the others that have been conducted since 1961, it will not surprise you to learn that nothing alien has been detected. After all, this would be a very different book if we had picked up any kind of anomalous signal. There have certainly been surprises along the way; yet, where unexpected signals have been discovered, each and every one has been demonstrated to be either of terrestrial origin or from a new class of astronomical source. To some scientists the silence heard from SETI searches is the deafening conclusion of a quest doomed to disappointment from the outset. To others, true believers, SETI searches are the only logical response to the question, Are we alone? The truth, as always, lies somewhere between these two opposing views, and we, as astrobiologists, must ask a different, perhaps more pertinent question: If no aliens have been detected up until now, how long must we continue to search before we hear something?

For example, can we extrapolate from the number of stars that have been unsuccessfully searched so far to the number that we must investigate in order to contemplate our first detection? The answer, of course, is no, we cannot. The best encouragement one could offer is that, if we search more stars at more frequencies, then we *might* detect something. But of course, then again, we might not. In that sense one could label the continued willingness of SETI scientists to search more stars at more frequencies as the triumph of hope—perhaps the ultimate romantic quest. However, the question of why look is much more than just an abstract point—in funding SETI to learn about communicating aliens, do we run the risk of denying funds to astrobiology projects that seek to learn about planets or moons in our own solar system that host less advanced life? These opposed philosophies led NASA to initiate its own SETI program in 1992 and then led the US Congress to abruptly cancel it just one year later. Projects such as Breakthrough Listen have happily rendered such debates moot by securing $100 million in private funding for ongoing SETI research—a number that, compared to many of the projects we have encountered on our astrobiological journey thus far, is significant only for how modest it appears.

We might contemplate the further question, What can we learn from an unresponsive universe? To consider this, we can write down a much simplified Drake equation, where the number of communicating alien civilizations is equal to the number of stars multiplied by the fraction that host broadcasting aliens ($N = N_{star} \times f_{aliens}$). Each star that we observe as part of a SETI search helps improve our knowledge of that fraction, whether it is transmitting or not. If, for example, we searched one million stars and detected no signals of alien origin, then we might well feel justified in stating that the fraction must be less than one in a million. Were we to observe more stars, still without success, we would improve our knowledge of the communicating fraction even further by limiting it to a smaller number.

Placing limits on any quantity of interest by counting the number of nondetections is a common and valid approach taken in many areas of fundamental science. However, in applying it to SETI before placing a limit on the broadcasting fraction, we would do well to consider whether our current efforts to detect alien signals from those star systems represent a comprehensive search or merely a partial one.

What if the aliens were indeed present on a planet orbiting one of the star systems we had observed yet happened to be broadcasting just outside of our radio window, or even in a completely different portion of the electromagnetic spectrum? What if an alien broadcaster considers a signal that flashes on and off once every ten thousand years to be an unmissable beacon? I mention these examples not to belittle those who perform SETI searches but merely to illustrate the daunting scale of the task involved in undertaking a truly comprehensive SETI search that would place strong limits on the fraction of communicating star systems out there in the event of a nondetection.

And daunting it certainly is. Though it would be difficult to overstate the importance to humanity if a signal were detected, the philosophy that underlies all SETI searches is essentially quixotic in that, to paraphrase the parting words of Cocconi and Morrison's paper, the rewards of success are considerable, but the odds of that success remain a profound unknown. To be fair, every scientist who undertakes a SETI search is fully aware of all these points, and their straightforward response demands respect: If they can undertake searches on their own dime, or at least on a threadbare, grant-funded budget, and if they don't get in the way of "respectable" scientific programs, then who are we to stop them?

We have so far talked a great deal about targeted searches for alien signals, but we should also consider the prospects for an accidental discovery, an unexpected signal in an astronomical observation unlike anything the cosmos has come up with before. There is of course a precedent that prompts this line of thinking. You wouldn't think that a four-acre sheep field in the Cambridgeshire countryside would be the most likely location for a Nobel Prize–winning astronomical discovery, but the facts of science can often be far stranger than fiction. This unassuming field hosted the interplanetary scintillation array, a collection of more than four thousand radio antennae, each the size and shape of a regular TV aerial (the sheep were an optional extra to keep the grass under control as a mower wouldn't fit between the rows of equipment). The array was originally constructed to study the twinkling of radio sources caused by the flow of the solar wind through interplanetary space. However, in 1967 Jocelyn Bell, then an astronomy graduate student at the nearby Mullard Radio Astronomy Observatory, used the array to record a series of regular pulses, every 1.3 seconds, that originated from a fixed astronomical point in

LGM-1 The interplanetary scintillation array in its four-acre field in Cambridgeshire. Jocelyn Bell, a graduate student at the time, detected a regular pulsed signal in the data collected using this array that was initially labeled LGM-1 (for Little Green Men). The signal was later confirmed as emission from a pulsar—a previously unknown class of rapidly rotating neutron star. The sheep used to keep the grass under control are out of the frame. Credit: Jocelyn Bell Burnell. Courtesy of the University of Cambridge, CC BY 4.0.

the sky. Originally labeled as LGM-1, for Little Green Men, the source of the pulses was quickly identified as a pulsar—a class of objects previously unknown in the universe—in this case a rapidly rotating neutron star produced in the blinding flash of a supernova explosion.

Although this episode did not mark the long-awaited moment of alien contact, it did demonstrate the fact that each time astronomers have used innovative technology to look at the night sky, they have discovered something new about the universe as a result. However, time and again the nature of that discovery has turned out to be surprisingly different from the one originally expected. It is a timely point to consider as, in the coming decade, two novel observatory projects in particular will see their first light and provide astronomers with impressive new views of the universe. These facilities will take the form of an extensive radio observatory known as the Square Kilometer Array and an optical telescope dedicated to mapping the night sky known as the Legacy Survey of Space and Time. Although these facilities are designed to make new astronomical discoveries, each represents a dramatically different way of looking at the universe compared to what has been done before, whether as a result of using a novel radio array or, more subtly, of monitoring the night sky with a regular cadence to pick up previously unknown variable sources. Each represents that magic combination of capabilities that will offer scientists a glimpse into previously unknown realms of our universe and presents new opportunities to those who search for beacons of intelligent life.

The Legacy Survey of Space and Time, known as the LSST, is a ten-year project scheduled to begin in 2025. Using an 8.4-meter diameter telescope hosted at the Vera C. Rubin Observatory in Chile, the LSST will survey the entire southern sky with an immense camera capable of capturing the light from an area equal to forty full moons. Yet it is the novel strategy of observing each patch of sky in a regular, repeating pattern—the shortest period being as little as every 3.7 days—that will provide astronomers with an unprecedented view of how the cosmos blinks, flickers, and flashes as a function of time. Astronomers like to say that telescopes provide a window on the universe, whether that window is characterized by the portion of the electromagnetic spectrum used to make those observations or the level of sensitivity afforded to view the very faintest of stars and galaxies. However, the window of time represents a truly novel aperture through which we have rarely observed the night sky in the manner afforded by the LSST. Although this telescope has been designed to perform astronomical surveys, no one is entirely sure what new signals will be discovered when we look at large portions of the night sky with such regularity. Given that your guess as to how distant aliens might signal their presence is as good as mine, this new window of time may well provide SETI searchers with a rich vein of data in which to prospect for the rare and precious signs of contact.

In contrast to the LSST, the Square Kilometer Array is a new multifrequency radio telescope spread across the radio-quiet territories of Australia and South Africa. In terms of collecting area—the hint is in the name—the facility is equivalent to a single telescope just over a kilometer in diameter. Given that the European Extremely Large Telescope we met in chapter 4 will feature a mirror only forty meters across, the Square Kilometer Array's name is remarkably modest and functional. It is an audacious project that aims to combine the radio signals received from thousands of antennae into an interferometric array. Not only will the array function as a single giant telescope, but the extended network of radio antennae will provide observations of exquisite spatial resolution and clarity. Although the SKA will not explore the time domain of the universe to the same extent as the LSST project, for scientists used to scouring the radio portion of the electromagnetic spectrum for intelligent signals, it represents a huge increase in listening power. Once completed in 2029, it will be some fifty times more sensitive than any predecessor. To appreciate that number, had the Extremely Large Telescope aimed for a similar jump in capability, it would have needed to weigh in with a truly colossal seventy-meter mirror.

As astrobiologists always on the lookout for the spurious, the weird, and the clues that just don't seem to fit, the novelty and innovation represented by the LSST and SKA should fill us with a sense of keen anticipation for the new and unexpected tales that the universe is about to tell us. Despite this, we are no still closer to knowing if we will discover an alien signal hidden deep within the petabytes of data that are expected from these formidable new observatories. At the same time, however, we must recognize that the possibility of detecting an alien broadcast exists and, perhaps more so, that the idea itself is plausible.

Yet what if the evidence for intelligent aliens lies right before us, and our Earth-based prejudices have caused us to overlook it? Might we be living our lives, such as we did before Antonie van Leeuwenhoek turned his microscope toward a drop of water, oblivious to the fact that new forms of life could be swimming right before our eyes? We have so far in this chapter considered the prospects for discovering advanced aliens based on what they *broadcast*. However, an alternative approach might be to search for what they *build*. Although the discovery of so-called technosignatures might sound far-fetched, that hasn't stopped a number of scientists from speculating on the strange nature of recent astronomical happenings.

Exhibit A in the case for technosignatures is Tabby's star, also known by the alias KIC (Kepler input catalog) 8462852. In 2015 astronomer Tabetha Boyajian and her colleagues alerted the scientific community to the strange dimming observed in this otherwise unremarkable star. That scientific publication was followed in 2017 by observations that showed rapid and sizable dips in the light coming from the star. The anomalous dimming of this star seemed to lie beyond what one would expect from a dusty protoplanetary disk or even more exotic explanations such as swarms of comets or the debris from colliding planets. Tabby's star was clearly an astronomical oddity and, as all such objects should be, was the target of follow-up observations from a number of observatories including the giant Keck telescopes on Hawaii, with even TESS sneaking in a few observations from its Earth-orbiting vantage point. The apparent failure of standard explanations to account for the brightness dips led some to speculate further on the nature of Tabby's star, and latent within the minds of science-fiction fans among the scientists who studied it lay the idea of a Dyson sphere, a megastructure constructed almost entirely around a star in order to capture as much energy as possible for a resource-hungry alien civilization. However, if that explanation strikes a chord of memory in your mind, then you might consider whether, in place of canals on Mars, were some in the scientific community seeing alien scaffolding around a distant star.

The attraction of using the idea of an alien structure to explain a confusing set of observations is twofold. First, there are few physical rules that tell you how they are built, so you are free to make them up in a way that perfectly matches the observations. Second, as an astronomer commenting on the likelihood of a vast alien space project, you are going to garner a flattering amount of attention from the media. What was true for Percival Lowell at the turn of the twentieth century remains true for the astronomers of today. The downside of favoring the exotic over the mundane to explain a fascinating stellar quirk is that even a nonscientist is able to answer the question of what is more likely—an alien megastructure or the possibility that we have more to learn about the properties of dust around young stars. To her credit, Tabetha Boyajian, the astronomer whose name the star bears, has always presented this object as a space oddity for which any claim as to its nature—astrophysical or otherwise—must be supported by primary evidence. And she has tolerated, with light-hearted grace, the claims that others have made for this enigmatic star.

Exhibit B in the case for technosignatures is no less strange, and it concerned an unexpected visitor to our solar system. In 2017, a bumper year for alien mysteries, the Pan-STARRS telescope at the summit of the Haleakala volcano on Maui detected an asteroid as it zoomed through the inner solar system, already beyond Earth and heading toward the outer planets. Further observations added additional points to our knowledge of its orbital trajectory, and it quickly became clear that this object had arrived from beyond our solar system and was streaking past the planets on its way back to interstellar space. Though astronomers were not afforded a close-up view of this asteroid, the variable scattering of sunlight from its surface suggested that it was a highly elongated cylinder some hundred meters long and that it displayed the dim red coloration characteristic of distant bodies from our own solar system. The object was named 'Oumuamua, a Hawaiian word that means "first distant messenger."

That an object could arrive in our solar system from across the wide sweep of interstellar space was not the mystery that vexed astronomers. 'Oumuamua was traveling at about twice the speed of our fastest space probes and would cover the distance between our solar system and the nearest stars in some fifty thousand years—the merest twinkle in the lifetime of a star or its planets. The nagging problem that seemed to defy explanation was its apparently extreme shape, unlike any other solar system body that we had previously encountered. At this point, as with a well-signposted movie trope, I think you can tell what is coming next.

In 2018 Harvard astronomer Avi Loeb and his collaborator published a paper in the *Astrophysical Journal* in which they suggested, among other ideas, that 'Oumuamua may have been an alien spaceship deliberately sent to our solar system on a reconnaissance mission. A generous take on this exotic suggestion might be that the authors were dropping some light speculation into an otherwise serious paper in order to emphasize the paucity of more reasonable explanations for this object. However, Loeb then followed up this suggestion with further articles, including a piece in *Scientific American*, that continued to advocate his alien hypothesis and has come to resemble a modern-day crusade that Percival Lowell, famed interpreter of Martian canals, would be proud of.

Sherlock Holmes counseled us that when you have eliminated the impossible, whatever remains, however improbable, must be the truth. Of course, key in this astronomical detective story is the question, Can we be sure that a strangely shaped but otherwise ordinary asteroid is impossible? Once again the answer is no. Impossible is a very strong word, and following the unsurprising frenzy of attention attracted by Loeb's claims, a number of scientific papers have pointed out that while 'Oumuamua represents an unusual object, it is by no means an impossible one.

None of this changes the fact that 'Oumuamua is an interstellar visitor, even if a purely natural one. Nor does it diminish the wonder of this unexpected and unusual event. For, if you really want to speculate, can we not instead imagine a future opportunity to recognize, intercept, and return samples from such cosmic interlopers as they pass through our solar system—a next-generation *Hayabusa* or *OSIRIS-REx* mission? Humans have long dreamed of visiting the distant stars, but how charmingly ironic is it to realize that their asteroidal emissaries have been visiting us all along.

If fate decrees that we do receive—and recognize—an alien message, what happens next? How would we react? At this point the questions we must answer have more to do with our own society than with any alien civilization. One point seems clear, though: Even an undeciphered message of alien origin would be a major event. Although the regularly flashing beacon of a lighthouse conveys no coded message, it would be wrong to conclude that its rhythmic pulse is devoid of meaning. Should an alien transmission be received, even if it contains no message apart from "We are here!" the reception alone of such a signal would clearly carry its own cosmic import. If one day we do indeed detect an alien signal that carries with it the unmistakable signature of intelligence, that would—with no need to translate it—represent an utterly profound discovery. In an instant it would change our perspective on our place in the cosmos. Unless Douglas Adams was unusually prescient when he suggested that a message might be a warning of our impending replacement by an interstellar bypass, the actual content of any message might be less important than the very fact of its reception.

Dare we broach the question of how receiving such a message would affect humanity? I use the word "dare" very carefully in this context. In one sense it is a question that is all too easy to answer—albeit with a confusing cacophony of responses. I suspect, however, that it is much harder to provide a single meaningful answer to the question of how alien contact would affect our society—mainly because, as with the search for life itself, we have little primary evidence upon which to base our answer. Perhaps, in the spirit in which this book is written, we could answer it using analogy, by reflecting on how we have reacted to startling new discoveries in our recent past. Although, if we consider how we have greeted the scientific evidence for the human causes of climate change, the efficacy of COVID-19 vaccines, or even moon landings, the answers we obtain from this approach might not be very palatable, as, in doing so, we would hold up an unforgiving mirror to our present societies.

But what might the decipherment of any message look like? During my own astrobiological journey I have detected a subtle yet pervasive assumption among scientists that the universal logic of mathematics will provide a master key to unlock the meaning of any alien signal we might happen to receive. Don't get me wrong—it is a plausible idea and one that marks a sensible starting point for any attempt to translate a message. However, I am sure that by this point you will have realized that if there is one thing I am wary of, it is the speculation that can result from received, though untested, wisdom.

Not surprisingly, this turns out to be another area where the pale blue data point has much to teach us. We share planet Earth with a host of fellow creatures. Some of them share our homes and emotional lives and possess a deep understanding of our human natures. Others have brains as large and complex as our own and presumably lead rich lives, however much they may be detached from ours. All life communicates, whether chemically or by using touch, visual signals, or—of most interest to us—sound, from at a distance. The fact that animals communicate is important in its own right. Communication doesn't need to include more complex ideas that we might call language in order to confer an evolutionary advantage to those that make use of it. But as astrobiologists in search of analogues on our pale blue data point that might teach us how to understand an alien language, clearly there are creatures that offer both a preeminent opportunity and a deeply profound challenge as we seek to encounter and potentially understand languages that are not our own.

In his groundbreaking television series *Life on Earth*, David Attenborough presented our species, *Homo sapiens*, as compulsive communicators, occupying one end of the continuum of life on Earth by virtue of our use of communication and language as a survival tool. Yet the term could equally well be applied to the many species of dolphins that swim in our oceans and communicate not just compulsively but in complex and varied ways that we are only just beginning to understand. Dolphins present an Earth-based analogue of what it might be like to encounter an alien language that is both compelling and confounding. Could we therefore take inspiration from Douglas Adams, whom we should surely consider an honorary member of the Order of the Dolphin, and eavesdrop on some of our nonhuman neighbors closer to home to get a sense of what forms alien communication might take? You may well think that this is a playful exercise. Yet it is worth considering that, if deciphering an alien message really lies within our capability, then working out what the dolphins have been going on about all this time should indeed be child's play in comparison. After all, we are separated only by some 95 million years of evolution—it's not as if they are aliens after all. But if it should turn out to be rather difficult for humans to understand dolphin communications, then we might receive a welcome lesson in humility as we contemplate the idea of decoding a truly alien signal.

My journey to find answers ultimately found me floating in the clear waters of the Bahamas as I explored one of the most engaging analogues of alien life that our pale blue data point can provide. I joined a team of scientists from the Wild Dolphin Project who have studied the Atlantic spotted dolphins of these shallow seas for nearly forty years, in a personal attempt to truly appreciate the challenge confronting those who hope one day to decipher an alien broadcast.

The research vessel *Stenella* offered a floating base for twelve crew members, researchers, and guests as we roamed the waters of the Bahamas in search of contacts. Every team member aboard had a role to play, and daily life centered on the hourly watches that each of us would stand on lookout for dolphins in the seas around us. Once spotted, the dolphins would often approach the twin prows of the sixty-two-foot catamaran in search of a free bow ride, and from there the scientists would snap a quick sequence of photos in order to identify and catalog individual dolphins. If the pod seemed interested in lingering, then we guests and the researchers would quickly equip ourselves with fins, snorkel masks, and video gear to join the dolphins in the water.

Everything would change upon entering their world, a threshold crossed in more ways than one. The effortless curve of a dolphin's passage inspires in us humans deep joy and an instant sense of connection as we float in an unfamiliar world, beguiled by the company of creatures that seem uncannily present in theirs. We are the visitors, and at times we would become the object of a keen and separate curiosity on the part of those we nominally came to study.

Although in some encounters we met calm, dozing pods or family groups contentedly foraging, in others we might encounter boisterous packs of young males out looking for interspecies trouble. Whatever the energy level of an encounter, all were characterized by nearly constant communication among the dolphins, and their vocal repertoire includes a seemingly unlimited range of whistles, clicks, and pulsed calls. Even the calmest of encounters with wild dolphins carry with them a charge of excitement. The water comes alive with their sonic chatter of squawks and whistles. A ripple of clicks passing through your body signals that a dolphin has taken an interest in you as an individual and has taken your measure in a burst of ultrasound. Taken together, the overall experience can be like trying to watch several overlapping TV stations at once—so much seems to happen so quickly that our brains struggle to keep up with mere cause and effect. My fellow researchers had advised me to spend my first few encounters just taking in the spectacle instead of attempting to apply a reasoned order to the dolphins' fast-forward lifestyle. Thus I allowed myself to hang suspended in the sunlit water like some neutrally buoyant astronaut as I watched two distantly related terrestrial species act out a compelling cosmic analogue.

Each encounter with dolphins proceeds, as far as is possible, on their terms and in their world. Each Atlantic spotted dolphin can be recognized as an individual from the unique pattern of speckles on its body, in addition to scratches, marks, or gouges on its fins. The scientists who study them week in and week out through the summer field season are akin to anthropologists studying remote human cultures by passively observing their daily lives. Whatever the encounter, every fleeting moment is observed by a snorkel diver carrying a handheld directional hydrophone and video recorder. For researchers attempting to navigate this intimidating rush of experience, the cardinal rule remains: Don't speak unless spoken to, and don't interact unless interacted with. The best encounter from a scientific standpoint is one in which the researchers remain spectators, recording the daily lives of others. However, sometimes the dolphins have other ideas and, for want of a better word, will simply hang out and play with the visiting humans until their interest in us is used up and, with a graceful arc of fin, they disappear to pastures new.

Only later, at the end of the day, can the accumulated recordings be patiently reviewed, the rushing cascade of information played back as a slow-motion trickle of acoustic events. Each squawk, chirp, and click can be tagged and matched not just with the signaling dolphin's own body posture but with the swift reactions of its companions. This is the point where dedicated and patient fieldwork in the water has its reward as patterns in the vocal and visual signals can be correlated with the context of the encounter—feeding, socializing, harassing, playing, mating—often all at once. The ability to slow everything down is key, as the direction and tone of the conversation seem to zigzag every quarter of a second or less.

Dolphins are communicators every bit as compulsive as we are—they seem to do nothing but. We can observe their vocalizations in context, but we haven't yet cracked the code of how the vocal parts fit into a meaningful whole. Human speech is based upon the arrangement of distinct units of sound—phonemes—that form the linguistic building blocks of our spoken language. Although we can observe dolphins using similar structural units in their vocalizations, just as in human speech, any meaning that they convey is likely to be contained in the myriad ways in which these acoustic building blocks can be combined together. To continue the analogy, in human speech phonemes are combined to form words, and ordered sequences of words are used to convey complex meaning.

The scientists who study dolphin vocalizations find themselves in a position like that of a young child first exposed to the rich and varied world of human sounds. We are just beginning to recognize phonemic order in the acoustic world of dolphins, and we are making our first tentative attempts—one that babies make look so easy—to associate groups of sounds with meaning. The analogy to human infants is apt because the one outstanding fact that we have learned so far from studying Atlantic spotted and bottlenose dolphins is that they refer to each other by names—individual signature whistles—that are learned in early childhood. Compared with the scale of deciphering a species-level system of communication, the discovery is modest. However, it is also deeply significant, hinting as it does, by virtue of the analogy with human names, at the existence of individual identity and a sense of self in a species other than ourselves.

Looking beyond this discovery, the challenge—and the hope—is that dolphins might be using vocalizations in a manner that is fundamentally different from the way in which humans use language. The challenge, of course, rather like the challenge of recognizing truly alien life, is to appreciate a different approach to information and meaning. The hope is that, in doing so, dolphins can offer us a terrestrial opportunity to broaden our horizons and, if only for a brief moment, step outside the confines of our experience as *Homo sapiens* and into a world beyond our own. After all, as an alternative to deciphering a nonexistent alien message, one could do a lot worse.

Our journey thus far has brought us to the outer limits of the Drake equation. We are swimming far from shore and are grateful when a passing dolphin offers us a helping flipper in our attempts to understand worlds of communication and expression far beyond our own. However, much as we have seen with almost of all of the terrestrial analogues we have explored so far, those glimpses beyond our Earth-bound human perspective are just that—mere glimpses, partial and incomplete. No visit to the rock outcrops of Western Australia or the depths of the Pacific Ocean can replace actually going to Mars or Europa in search of brave new worlds. Encounters with our fellow creatures from the animal kingdom, even though desperately important as we seek to understand and protect our natural world, can only serve as practice questions while we wait for the real test of our communication skills to arrive over the galactic airwaves.

As our journey nears its conclusion, it is perhaps time to consider where the experiences we have assembled thus far leave us as scientists. How can we, without ever having discovered the merest cellular speck of evidence for alien life, call ourselves astrobiologists? Can a collection of tangential ideas, drawn from all points of the scientific compass, be made to fit into the circle of a new discipline? Perhaps the answer is yes. However, to achieve this, a great many scientists have had to be brave enough to step beyond the confines of previously separate research fields into the uncertain no-man's-land that divides them. And while I believe that there does exist such a thing as an astrobiologist, you might be hard-pressed to get one to admit to it. For the truth is that many of the scientists who work in fields related to astrobiology are reluctant to label themselves as astrobiologists. Partly this is professional modesty, as being an expert on terrestrial oceans does not necessarily make one an expert on Europan or Enceladan ones, let alone any life they might contain. However, as I hope we have learned from the stories in this book, it doesn't make one a complete newbie either. Perhaps that reticence comes from a form of impostor syndrome, as if discovering alien life is the minimum entry fee to an exclusive club. Yet the secret to being an astrobiologist, as in all other fields of scientific research, is to have the confidence to apply what you do know to answer questions that truly matter—all the while remaining aware that whatever knowledge you do possess is guaranteed to be incomplete, likely misleading, and possibly even wrong. Given that the odds appear to be stacked against us, who might be brave enough to even try?

Astrobiology has always been interdisciplinary—one just has to look at the backgrounds of the original members of the Order of the Dolphin to be reminded of that fact. But the outlook of astrobiology has broadened considerably in the sixty-plus years since Green Bank and the Drake equation. So who are the present-day astrobiologists? In this book I have shone a light on several areas of Earth-based or remote space research that I think either offer promising analogues of alien habitats or actually explore biologically compelling environments beyond Earth. The fact that I shine that light more closely on the individuals, groups, and research teams I have been fortunate enough to partner with in the field or laboratory is due to this being an effective way to observe astrobiology in action. Yet it is also an unfair one, as the restrictions of writing what I hope is a focused and readable book has meant that I've left out many wonderful stories from other scientists. I have attempted to capture a snapshot of a scientific discipline at one moment in its developing story. The picture is incomplete, necessarily so, as it would be impractical to include every nook and cranny under scrutiny in a wide-ranging field. Necessary but also deeply unfortunate. Who is to say in what overlooked recess of the solar system we might discover a cryptic yet flourishing habitat? Which niche of terrestrial existence might reveal the critical biological clue that has so far been passed over unnoticed?

A final shortcoming (apologies really are due at this point) is that any snapshot is also incomplete in time. Although I enjoy speculating upon what revelations an astrobiology book written one hundred years from now will contain, I also wince at the discoveries that will be made in the coming year or two that I will be unable to squeeze into the final few pages of this one. The samples returned by *OSIRIS-REx* will be analyzed and reveal their asteroidal secrets. The Extremely Large Telescope will deploy its immense, unblinking gaze on the small handful of habitable, rocky worlds that we currently know. A priceless cache of rock samples will be returned to Earth from Mars, and *Europa Clipper* will probe the mysteries of the Europan ice sheet. I remain certain that there will indeed be new discoveries, as even the limited snapshot I have been able to provide here reveals the telltale blur of motion that speaks of the rapid pace of present-day exploration.

The themes I have focused on in this book are those that I am confident will bring about new discoveries and new knowledge. Rocks will be returned from Mars; we will probe, at least tentatively, the ice sheets of Europa; we will characterize the major atmospheric constituents of exoplanet atmospheres; and we will hold in our hands further samples of solar system asteroids. We will make new discoveries and expand the horizons of our knowledge—of that I am sure. But will we discover new life? On that point I am far less certain. I could be bold and claim that any one of the techniques I have described in this book is likely to discover alien life in the coming decades. Were I to do so, I would be selling you easy, empty words. The more compelling statement is that, as far as we are aware, we are looking in the right directions, and we need to keep doing so. That claim should not be taken as overconfidence—as I welcome the possibility that the universe has some further surprises in store for us. Yet perhaps from this perspective we can at last appreciate that little bit more clearly the oft-quoted parting words of Cocconi and Morrison, who told us that in our astrobiological quest "the probability of success is difficult to estimate; but if we never search, the chance of success is zero."

The original members of the Order of the Dolphin who met at Green Bank Observatory in 1961 were the first people on Earth to see the Drake equation. Being curious individuals, they quite naturally attempted to assign values to the conspicuously unknown terms in the equation. Many of their guesses were speculative—one might even say optimistic. However, the scientists themselves may well have responded that in making those assumptions they were seeking ballpark answers that might offer an approximate sense one way or another of whether they were dealing with a hypothetical Milky Way galaxy brimming over with intelligent alien life or one devoid of galactic neighbors (not to mention hope).

As we have seen for ourselves, the fraction of stars hosting at least one planet turns out to be high, encouragingly so if you seek to proceed further along the succeeding terms in the Drake equation. Based upon their admittedly limited experience of life on Earth, the members of the Order of the Dolphin saw only further encouragement as they considered the later terms along the path to intelligent, communicating life. Consequently, they assigned values to these terms that were all close to one. As a result of these guessed-at values, the Drake equation could be rewritten in a simpler form—one might call it the optimistic limit of the equation—where $N = L$. What this statement means is the that the number of communicating civilizations in the Milky Way may well be very close to the number of years over which such civilizations exist and transmit.

Even Frank Drake was cautious of overinterpreting these attempts to coax hard numbers out of his equation, wary of treating it like some cosmic computer capable of spitting out deep philosophical truths. His view was that the Drake equation offers a valuable summary of the factors that play a role in answering the question of whether anyone is out there. Although it provides a sobering reminder of just how much we do not currently know, that lack of knowledge doesn't make it any less relevant. However, I do think that the statement that $N = L$ is also potentially quite profound. If taken at face value, one can think of it as informing us that, should we wish to increase our chances of detecting intelligent alien signals, then our continued existence on this pale blue data point represents the only number that we control—everything else is decided for us by the universe. Every year that we add to this number, every year that we tackle climate change, that we reduce the threat of war, that we learn to live as careful tenants on a fragile planet, we increase our chances to discover alien life in the universe—assuming, of course, that we continue to search.

Carl Sagan probably understood this well before he saw the image of the pale blue dot in 1990. For many of the generation that grew up during the golden years of the space age it was the image of Earth rising over the rim of the moon, taken on Christmas Eve 1968 by *Apollo 8* astronaut William Anders, that provided the first, vulnerable perspective of Earth as viewed from space. Once Sagan saw the later image of Earth suspended in a sunbeam, it perhaps only reinforced his long-held views on our cosmic context.

So what might our future be? If we can overcome the present challenges we face as supposedly intelligent, communicating creatures living on our home planet, might this imply that other, alien races could also do so? However, we should also consider the converse of this argument—all the more so for its lack of appeal. We have so far focused on the fraction of planets on which life proceeds to the next stage of the Drake equation. But what of those planets where life fails to make the cut, the fraction that instead falter at the next step along the road? Might a sudden drop-off in the numbers assigned to each symbol in the Drake equation mask unguessed-at cosmic extinctions? This idea has been cast before, most notably by the economist and futurist Robin Hanson, as the great filter. Perhaps the journey of life from basic to complex and ultimately to that of a communicating (or even a traveling) civilization is not as easy and straightforward as we might like to think it is. Were we to discover evidence of basic life in the solar system, or even on an exoplanet, yet found that the galaxy as a whole remained stubbornly silent on the subject of communicating aliens, would this imply that the great challenges to existence lay not in our past but in our future? At what point in our own journey might we encounter that great filter, the winnower of existence, where the chaff of unenlightened, or merely unfortunate, planetary civilizations is separated from the wheat of long-lived alien races?

At present we face a series of crises. Human actions are changing the climate of planet Earth. The weight of primary evidence can be supported by no other conclusion. The open question is what will be the consequences of this change. We are also consuming our planet's resources at a rate at which depletion—of forests, of fresh water, of clean air—is a desolate possible future. Are these merely the challenges of our time, the growing pains of an adolescent planetary species finding its place within the cosmos? If so, who can guess what challenges might face a more advanced civilization? One might be tempted to hope it gets easier, although, having reached this far in the game of life, it would be unwise to adopt a false sense of security.

The passing of the year 2022 took with it two notable figures from our story of astrobiology—James Lovelock and Frank Drake. Lovelock gave voice to the idea that our planet represented a complete biological unit in its own right—a self-regulating, ordered organism that carried with it its own recognizable signatures of life. Drake showed us a path to the distant, living worlds that many have speculated exist beyond our own. Whether that path leads only to basic life in our solar system or ultimately to new civilizations among the stars is perhaps irrelevant at this early stage in our search.

James Lovelock and Frank Drake did not witness the discovery of alien life in their lifetimes, and, perhaps, nor did they expect to. However, at this point in our journey together, I hope that you now have a very personal sense of that moment approaching. Earlier in this chapter I posed the question of how we might greet the discovery of alien life. Having talked about the challenges we now face, my

Outermost house Henry Beston, author of *The Outermost House*, lived alone in this tiny shack on the Cape Cod coast from 1926 to 1927 as he observed the ebb and flow of the natural world around him. He never saw our modern view of Earth from space, yet he did not need to: His view of life on Earth was as total and complete as any perspective we have today. That perspective has taught us that Earth is unlikely to represent a single, pale blue dot of life in the universe and that we, as its inhabitants, should treasure the privilege of existing on it. Credit: Courtesy of the George. J. Mitchell Department of Special Collections & Archives, Bowdoin College Library, Brunswick, Maine.

only response is to offer words of hope, and even these are not my own. They belong to the American naturalist and writer Henry Beston, who in 1928 summed up in his book *The Outermost House* the feeling that resulted from his own unobstructed encounters with nature. From his modest shack at the edge of the Atlantic Ocean, he never saw the view of the pale blue data point that we have been privileged to see. However, like many of those who are truly sensitive to the world around them, he didn't need to in order to grasp the elusive totality of life on our planet. Of the life he had encountered he wrote,

In a world older and more complete than ours, they move finished and complete, gifted with the extension of the senses we have lost or never attained, living by voices we shall never hear. They are not brethren, they are not underlings: they are other nations, caught with ourselves in the net of life and time, fellow prisoners of the splendour and travail of the earth.

From this final vantage point, I can't think of any other words that better describe the sum of our astrobiological encounters with the living organisms of planet Earth, whether they be with silent stromatolites, abyssal tube worms, or even a wild dolphin in the bright waters of the Bahamas. Yet, in a last parting gift, a final analogy from our pale blue data point, those words may also serve as our guide as we continue our quest to discover alien lives.

ACKNOWLEDGMENTS

I am deeply grateful to the team at the University of Chicago Press for their help at every stage of the journey toward completing this book. Particular thanks are due to my editor at Chicago, Joseph Calamia, who has been an enthusiastic champion of the book every step of the way. I also wish to express my gratitude to the anonymous reviewers who took the time to cheer, correct, and guide the book along the route.

I will always be indebted to those scientists, engineers, and explorers who helped me to realize the journeys undertaken in this book.

Thank you to Megan Cook, Allison Fundis, and the crew at the Ocean Exploration Trust and on board the *Nautilus* for the chance to join a world-class team of explorers.

Thanks to Martin van Kranendonk for the advice and encouragement to visit the fossil sites of early life on Earth and to Matt Kilburn for welcoming me to the Centre for Microscopy, Characterisation and Analysis at the University of Western Australia. Thanks also to Vivian Sun at NASA's Jet Propulsion Laboratory for guiding me through Mars rover operations and the day-to-day life of a Mars scientist.

Gordon Walker and Stephenson Yang, both at the University of Victoria, were generous with both their time and their memories of their pioneering research into exoplanets.

Thank you to Elizabeth, Steve, and Wolf at Meteorite Adventures for taking me under their wing and sharing their knowledge and experience with me in the field. Thanks also to Chris Cokinos for advice and encouragement that ultimately saw me hunt meteorites for myself.

It was a privilege to be allowed to join Denise Herzing and the team at the Wild Dolphin Project, and thanks also to Kathy Reynolds for giving me the chance to join the scientists in the Bahamas for a second time.

Final thanks go to my wife Sara and our child Osiris for their encouragement and love that made this book possible.

FURTHER TRAVELS

Many learned books contain a section that offers readers the chance to expand upon their erudition by reading even more books and articles. However, having traveled across planet Earth in search of answers to some of astrobiology's most pressing questions, I also wanted to offer you, the reader, the chance to follow in my own footsteps—whether metaphorically or indeed literally. I hope the following list might serve as your guide.

CHAPTER 1

Carl Sagan and Ann Druyan muse further on our cosmic perspective in *Pale Blue Dot: A Vision of the Human Future in Space* (Ballantine Books, 1997); and should you wish to get as close as possible to the *Voyager* spacecraft without leaving Earth, you can visit an engineering copy of the probe on display at the National Air and Space Museum in Washington, DC.

The Extraterrestrial Life Debate, Antiquity to 1915: A Source Book, edited by Michael J. Crowe (University of Notre Dame Press, 2008), is a comprehensive reference work that presents the original texts of many of the premodern thinkers of astrobiology and allows you to read the words of Bruno, Kepler, and Lowell as they wrote them.

The original scientific papers that describe the *Viking* biology experiments, the *Galileo* Earth flyby results, and NASA analyses of ALH84001 are: Gilbert V. Levin and Patricia Ann Straat, "Completion of the Viking Labeled Release Experiment on Mars," *Journal of Molecular Evolution* 14 (1979): 167–83; Carl Sagan, W. Reid Thompson, Robert Carlson, Donald Gurnett, and Charles Hord, "A Search for Life on Earth from the Galileo Spacecraft," *Nature* 365, no. 6448 (1993): 715–21; and David S. McKay, Everett K. Gibson Jr., Kathie L. Thomas-Keprta, Hojatollah Vali, Christopher S. Romanek, Simon J. Clemett, Xavier D. F. Chillier, Claude R. Maechling, and Richard N. Zare, "Search for Past Life on Mars: Possible Relic Biogenic Activity in Martian Meteorite ALH84001," *Science* 273, no. 5277 (1996): 924–30.

While we are on the subject of ALH84001, you might also take a look at Kathy Sawyer's *The Rock from Mars: A Detective Story on Two Planets* (Random House, 2006) for a gripping behind-the-scenes story of the most famous Martian meteorite of all.

Frank Drake and Dava Sobel, *Is Anyone Out There? The Scientific Search for Extraterrestrial Intelligence* (Delta, 1994), tells the wonderful story of Frank Drake's journeys through astronomy.

CHAPTER 2

If you want to go exploring with the E/V *Nautilus*, then you should navigate your way to nautiluslive.org. Here you can watch the live video feed and listen to scientists and engineers at work. If you are a scientist, educator, or student and you want to join the team on board the ship as a crew member, then follow the "Join" tab from the main page.

I joined the *Nautilus* as a science communication fellow and shipped out of Victoria, BC, to spend three weeks aboard supporting Ocean Networks Canada's NEPTUNE observatory in the northeast Pacific Ocean (see https://www.oceannetworks.ca/).

For an introduction to plate tectonics, one could not do better than to read Naomi Oreskes's wonderfully edited book, *Plate Tectonics: An Insider's History of the Modern Theory of the Earth* (CRC Press, 2003), where you can read the very human stories behind one of science's great paradigm shifts.

For a wonderful biography of Marie Tharp, a lesser-known but crucial contributor to our modern view of planet Earth, read Hali Felt's *Soundings: The Story of the Remarkable Woman Who Mapped the Ocean Floor* (Henry Holt, 2012).

Here is the journal article announcing the discovery of subsurface oceans on the moons Europa and Callisto: K. K. Khurana, M. G. Kivelson, D. J. Stevenson, G. Schubert, C. T. Russell, R. J. Walker, and C. Polanskey, "Induced Magnetic Fields as Evidence for Subsurface Oceans in Europa and Callisto," *Nature* 395, no. 6704 (1998): 777–80.

The academic paper describing the discovery of active geysers on Enceladus is C. J. Hansen, L. Esposito, A. I. F. Stewart, J. Colwell, A. Hendrix, W. Pryor, D. Shemansky, and R. West, "Enceladus' Water Vapor Plume," *Science* 311 (2006): 1422–25.

Holger Jannasch and Carl Wirsen's wonderful paper on exploring the alien ecosystems of deep ocean hydrothermal vents is "Chemosynthetic Primary Production at East Pacific Sea Floor Spreading Centers," *Bioscience* 29, no. 10 (1979): 592–98.

CHAPTER 3

My journey to the ancient fossil sites of Western Australia was made possible by Martin van Kranendonk and Jean Johnston, *Discovery Trails to Early Earth: A Traveller's Guide to the East Pilbara of Western Australia* (Geological Survey of Western Australia, 2009). I traveled along the authors' Trail 1 from Port Hedland to Marble Bar and then Trail 5 from Marble Bar to Nullagine. I also traveled to see living stromatolites at Hamelin Pool Marine Nature Reserve (see https://exploreparks.dbca.wa.gov.au/park/hamelin-pool-marine-nature-reserve for more details).

For a taste of a spicy academic debate, take a look at these back-to-back letters to *Nature* for a perspective on the scientific debates that occur at the cutting edge of such research: William Schopf et al., "Laser-Raman Imagery of Earth's Earliest Fossils," and Martin Brasier et al., "Questioning the Evidence for Earth's Earliest Fossils," *Nature* 416 (March 7, 2002).

By far the best way to keep up to date with all things to do with the Mars rover *Perseverance* is to use NASA's dedicated website, https://science.nasa.gov/mission/mars-2020-perseverance/.

CHAPTER 4

I worked at La Silla Observatory from 2002 to 2003, accumulating one hundred nights at the New Technology Telescope. Members of the public are welcome to go online and book free (actual) guided tours of La Silla at https://www.eso.org/public/about-eso/visitors/lasilla/.

Dava Sobel's excellent book *The Glass Universe: How the Ladies of the Harvard Observatory Took the Measure of the Stars* (Viking, 2016), tells the story of many of the pioneers in the field of astronomical spectroscopy.

Kaj Strand's planet detection claims are detailed in "61 Cygni as a Triple System," *Publications of the Astronomical Society of the Pacific* 55, no. 322 (1943): 29–32.

On slightly firmer ground are the claims of Bruce Campbell, Gordon A. H. Walker, and Stevenson Yang, "A Search for Substellar Companions to Solar-Type Stars," *Astrophysical Journal, Part 1* (ISSN 0004-637X) 331 (August 15, 1988): 902–21.

The zombie planets of Aleksander Wolszczan and Dail Frail are described in "A Planetary System Around the Millisecond Pulsar PSR1257+ 12," *Nature* 355, no. 6356 (1992): 145–47.

Here is the journal reference for the academic paper announcing the discovery of the first extrasolar planet, in this case orbiting the star 51 Pegasi: Michel Mayor and Didier Queloz, "A Jupiter-Mass Companion to a Solar-Type Star," *Nature* 378, no. 6555 (1995): 355–59.

The TRAPPIST-1 system is described in Michaël Gillon, Emmanuël Jehin, Susan M. Lederer, Laetitia Delrez, Julien de Wit, Artem Burdanov, Valérie Van Grootel, et al., "Temperate Earth-Sized Planets Transiting a Nearby Ultracool Dwarf Star," *Nature* 533, no. 7602 (2016): 221–24.

And here is the latest on the atmosphere of K2-18b: Nikku Madhusudhan, Subhajit Sarkar, Savvas Constantinou, Måns Holmberg, Anjali A. A. Piette, and Julianne I. Moses, "Carbon-Bearing Molecules in a Possible Hycean Atmosphere," *Astrophysical Journal Letters* 956, no. 1 (2023): L13.

CHAPTER 5

Although Primo Levi's *The Periodic Table* (Penguin Modern Classics, 2000) is an autobiography rather than a science book, Levi will nonetheless take you on an unforgettable human journey.

My own meteorite adventures were made possible by Meteorite Hunting Adventures, based out of Tucson, Arizona. Their website is www.meteoriteadventures.com, and they list upcoming expeditions to Morocco, Chile, and the US Southwest. Should you wish to read about a much fuller set of meteorite adventures than my own, including a very moving stint in Antarctica, read Christopher Cokinos, *The Fallen Sky: An Intimate History of Shooting Stars* (Tarcher, 2009).

I am often asked where one can purchase meteorites, and for a fair chunk of my collection I have used Aerolite meteorites (aerolite.org).

The first of what will be a series of results from the *Hayabusa2* mission are contained in Tomoki Nakamura, M. Matsumoto, K. Amano, Y. Enokido, M. E. Zolensky, T. Mikouchi, H. Genda, et al., "Formation and Evolution of Carbonaceous Asteroid

Ryugu: Direct Evidence from Returned Samples," *Science* 379, no. 6634 (2022).

Our ideas on the habitability of the early Earth were turned upside down by Simon A. Wilde, John W. Valley, William H. Peck, and Colin M. Graham, "Evidence from Detrital Zircons for the Existence of Continental Crust and Oceans on the Earth 4.4 Gyr Ago," *Nature* 409, no. 6817 (2001): 175–78.

CHAPTER 6

My journeys with dolphin researchers in the Bahamas were made possible by Dr. Denise Herzing and the Wild Dolphin Project, based out of West Palm Beach, Florida (wilddolphinproject.org). Their research cruises to the Bahamas typically last nine days, including two travel days. The Wild Dolphin Project has monitored the Atlantic spotted dolphins of the Bahamas banks for forty years, and joining them offers unprecedented access to a world-class research team and some quite exceptional dolphins.

Here is where SETI began: G. Cocconi and P. Morrison, "Searching for Interstellar Communications," *Nature* 184 (1959): 844–46.

For a cosmic whodunit you could do worse than to read Tabetha S. Boyajian, D. M. LaCourse, S. A. Rappaport, D. Fabrycky, D. A. Fischer, Davide Gandolfi, G. M. Kennedy, et al., "Planet Hunters IX. KIC 8462852—Where's the Flux?," *Monthly Notices of the Royal Astronomical Society* 457, no. 4 (2016): 3988–4004.

And if you are ready for science of a more speculative nature, try Shmuel Bialy and Abraham Loeb, "Could Solar Radiation Pressure Explain 'Oumuamua's Peculiar Acceleration?," *Astrophysical Journal Letters* 868, no. 1 (2018): L1.

The Outermost House: A Year of Life on the Great Beach of Cape Cod, by Henry Beston (Holt Paperbacks, 2003), provides a fitting end to this list and, perhaps, the start of your next journey.

INDEX

Page numbers in italics refer to figures.

absorption lines, 120, 125, 142
Abyssal Zone, 55
Acasta River, 175
accretion disks, 147
Acoustically Navigated Geophysical Underwater System (ANGUS), 42
Adams, Douglas, 190, 209, 211
adenosine triphosphate (ATP), 14, 56
AGB (asymptotic giant branch) stars, 154–55
Aldrin, Buzz, 25
ALH84001, 26–28, *26*, 94
All These Worlds Are Yours (Willis), 9, 65
Allende meteorite, 156–57, *157*
ALMA. *See* Atacama Large Millimeter Array (ALMA)
Alvin (submersible), 50–51, 58
American Association for the Advancement of Science, 89

amino acids, 3, 178
ammonia, 173
Amundsen, Roald, 70
Anders, William, 221
ANGUS (Acoustically Navigated Geophysical Underwater System), 42
Antarctica, 24–25, 69–70, 161
Apex chert, 90–92
Apollo program, 25, 141, 166, 221
aragonite, 83
Archean eon, 87, 90–92, 94, 108
Arecibo radio telescope, 127
Argentina, 181
Argus (ROV): camera of, 49; *Nautilus* and, 36, 46, 49, 100; oceans and, 36, 45, 48–49, 74, 100
Ariane 5 rocket, 148
Armstrong, Neil, 25
asteroids: amino acids and, 3; Bennu, 165, 172, 179, 189;

asteroids (*cont.*)
dust particles and, 167; as emissaries, 209; fossils and, 79; gravity and, 105; meteorites and, 7–8, 32, 155–60, 164–74, 177–81, 189; *OSIRIS-REx* and, 105, 165, 172, 178, 209, 218; 'Oumuamua, 207–9; Pan-STARRS and, 207; raw chemicals of, 7; Ryugu, 165–71, *167*, 179, 189; samples of, 8, 32, 105, 164–72, 178, 189, 209, 218; *Stardust* probe and, 165; Vesta, 164

astrobiology: ALH84001 and, *26*, 94; amino acids and, 3, 178; ATP and, 14, 56; biogenicity and, 94; biomarkers, 17–19, 137–38, 141–42, 149–50; carbon and, 3, 14–15, 23–25; chemistry and, 3–4, 7–8, 14, 18–24, 28, 32, 210; communication and, 190–95, 199, 205, 210–11, 216–19, 222, 224; definitions of life and, 20–22; DNA and, 2, 14, 22, 191; Drake and, 29–34, 190–92, 194, 217, 222; exoplanets and, 112, 115, 128, 131–33, 136, 143, 146, 148; Gaia hypothesis and, 18, 144–46; *Galileo* and, 15–19, 39, 66; goal of modern, 32; "Goldilocks zone" and, 113, 128, 130, 133, 136, 148; habitability and, 7, 104, 113, 134–36, 146; hydrogen and, 14; hydrothermal vents and, 5, 15, 29, 32; Jupiter and, 2, 5, 34; Lowell and, 22–23, 206, 208; Mars and, 3, 5–6, 8, 19, 22–29, 32, 34, 83, 91, 98, 218; measuring life and, 19–20; metabolism and, 3, 17, 20–23, 27; meteorites and, 176–79, 189; methane and, 16–18, 80, 137, 142, 147, 151; microbes and, 17–18, 23, 26–27, 29; Milky Way and, 12, 30, 32–33; minerals and, 2, 26–28; motivations for, 10–13; oceans and, 39, 45, 57–59, 63–64, 66, 71, 74; optimistic view for, 3; Order of the Dolphin and, 8, 196, 211, 217, 219–20; oxygen and, 3, 14, 16–17; photosynthesis and, 3; Project Ozma and, 192–98; proof of life and, 27–28, 94; RNA and, 59; ROVs and, 5–6, 19; "Sagan criteria" and, 17; samples and, 19–25, 32; Saturn and, 2, 5, 9, 32, 34; search criteria for, 2; SETI and, 191, 194–201, 204; stromatolites and, 83; Venus and, 9

Astrobiology Institute, 28
astronomical units, 122
Astrophysical Journal, 208
astrophysics, 118–21, 207
asymptotic giant branch (AGB) stars, 154–55
Atacama Large Millimeter Array (ALMA): abilities of, 182–83; Chile and, 114, 173, 182–87; L Tau and, 183–86; protoplanetary disks and, 183–88; stars and, 173, 183–84, 186
Atchley, Dana, 194
Atlantis space shuttle, 16
atmospheres: AGB stars and, 154; biomarkers and, 17–18, 138, 141–42, 149; Earth, 9, 16–18, 55, 82, 85, 135, 141–44, 147–48, 158, 170, 177, 180, 182; Europa, 67; exoplanets and, 3, 110, 125, 129, 135–44, 147–51, 218; *Galileo* and, 16–18; habitability and,

INDEX [235]

17–18, 85–86, 104, 135–38, 141, 143–44, 147–50; iron sulfide and, 53; lightning and, 17; Mars, 67, 85–86, 104, 218; meteorites and, 159, 164, 170; methane and, 16–18, 80, 137, 142, 147, 151; oxygen and, 16–17, 80, 82, 87, 137, 142, 151; pressure and, 55; silicon carbide and, 158; temperature and, 55, 129, 135–36, 177; Venus, 9, 67
ATP (adenosine triphosphate), 14, 56
Attenborough, David, 211
Australia: Apex chert, 90–92; Archean eon and, 87, 90–92, 94, 108; communication and, 204; exposed crustal rocks of, 87; fossils and, 6, 25, 32, 75–77, 81, 84, 87–93, 108, 216; *Hayabusa2* and, 170; iron and, 75–76, 80, 88; Marble Bar, 77–78, 80, 90; meteorites and, 156; Murchison, 156; North Pole Dome, 90; Pilbara, 76–77, 87, 90, 109; radio telescopes and, 204; rocks and, 6, 25, 41–44, 50–53, 75, 77, 79, 84, 87–93, 108, 156, 216; Strelley Pool Chert, 88, 90; stromatolites and, 75–76, 81, 88; temperatures of, 77
autonomous underwater vehicle (AUV), 70–71
autotrophs, 55–57, 63

bacteria: cyanobacteria, 82, 89–90; fossils and, 75, 80–82, 88–90; geology and, 37, 89; minerals and, 27; oceans and, 37, 54, 60–64; oxygen and, 60, 64, 82
Bahamas, 8, 212, 224

Ballard, Robert, 45
Barghoorn, Elso, 89–90
BARS (benthic and resistivity sensor), 51–52, *52*
Bassham, James, 195
Bell, Jocelyn, 201–2, *202*
Bennu, 165, 172, 179, 189
Benson, Andrew, 195
benthic and resistivity sensor (BARS), 51–52, *52*
Berlin Observatory, 117
Beston, Henry, 223–24
Beta Pictoris, 173
Big Bang, 188
Big Dipper, 120
binary stars, 113, 120–23
biogenicity, 27–28, 94
biomarkers: astrobiology and, 17–19, 137–38, 141–42, 149–50; atmospheric, 17–18, 138, 141–42, 149; exoplanets and, 137–38, 141–42, 149–50
black holes, 147
Blumberg, Baruch S., 28
Bolivia, 181
Boyajian, Tabetha, 205–7
Brasier, Martin, 91–94
Breakthrough Listen, 198
Bruno, Giordano, 11, 115, 149
Bunsen, Robert, 119
Butler, Paul, 129

calcium-aluminum inclusions, 156, 186–87
calcium carbonate, 83, 153
Calvin, Melvin, 195–96
Calvin cycle, 56
Cambrian explosion, 89
cameras: ANGUS and, 42; *Argus* and, 49; *Cassini* and, 38; digital, 124; *Hayabusa2* and, 169; *HIBOU*

cameras (cont.)
and, 166; Legacy Survey of Space and Time (LSST) and, 203; LIDAR and, 68; Mars Reconnaissance Orbiter (MRO) and, 97; meteorites and, 164, 166; METIS and, 141; OWL and, 166; Perseverance and, 99, 101–3; spectra and, 132, 141; telescopes and, 6, 111, 124, 132, 146, 203; TESS and, 146; thermal imaging, 61

Campbell, Bruce, 125–26

Campo de Cielo (Field of Heaven) meteorite, 159–60

Canada, 45, 49, 73, 89, 147, 175

carbon: astrobiology and, 3, 14–15, 23–25; Brasier on, 91; calcium carbonate, 83, 153; chemistry of, 153–54; chondrite and, 156–59, 165, 170, 174, 177–80, 186; exoplanets and, 147; formation from helium, 154; fossils and, 80, 82–83, 91–92; journey of, 153–56; meteorites and, 25–26, 153–60, 165–66, 170–74, 177–81, 186; oceans and, 55–60, 73; photosynthesis and, 55–56, 153, 195; protoplanetary disks and, 155, 173; silicon carbide, 154–55, 157–58

carbon dioxide: atmospheric color of, 80; exoplanets and, 147; photosynthesis and, 55–57, 62, 64, 82, 89, 153, 195; protoplanetary disks and, 173; radioactivity and, 58

Cassini space probe, 15, 38–39

Cerro Armazones, 139–40, 142

Challenger Deep, 44

chemistry: astrobiology and, 3–4, 7–8, 14, 18–24, 28, 32, 210; ATP and, 14, 56; Calvin and, 195; carbon, 153–54; chondrites, 156, 165, 177, 180; DNA, 2, 14, 22, 191; Earth's crust and, 43; exoplanets and, 119, 137–38, 143–44; fossils and, 82–83, 108; hydrogen, 14, 56–57, 60, 154, 173–74, 193, 197; isotopes, 108; Levi on, 153–54; meteorites and, 153–57, 164–65, 168–81, 188; methane, 16–18, 80, 137, 142, 147, 151; Miller–Urey experiment and, 180; nitrogen, 3, 14, 80, 92; oceans and, 37, 43, 50–51, 54–58, 62, 64, 66; oxygen, 82 (*see also* oxygen); photosynthesis and, 55–57, 62, 64, 82, 89, 153, 195; RNA, 59; sugars, 55, 57, 60, 82; sulfur, 9, 14, 60, 82–83, 92

chemolithoautotrophs, 57

chemosynthesis, 57–61, 64

"Chemosynthetic Primary Production at East Pacific Sea Floor Spreading Centers" (Jannasch and Wirsen), 57–58

chert, 79, 88, 90

Chile: ALMA, 114, 173, 182–87; astronomy done in, 7, 31; Cerro Armazones, 139–40, 142; European Southern Observatory and, 139; exoplanet exploration and, 110, 113–15, 139–40, 142; Extremely Large Telescope (ELT), 139–42, 148–49, 183, 204, 218; La Silla Observatory, 110, 114, 132–33, 138, 150; Legacy Survey of Space and Time (LSST), 203–4; meteorites and, 173, 181; New Technology Telescope (NTT), 114–15; Vera C. Rubin Observatory, 203–4

chlorophyll, 17, 56, 82

chondrite: chemistry of, 156, 165, 177, 180; LL 3.8, 159; meteorites and, 156–59, 165, 170, 174, 177–80, 186; NWA 6135, 159; Ryugu and, 170–71
chromatography, 23
Churyumov–Gerasimenko (comet), 67–68
clams, 42, 57
Clarke, Arthur C., 65, 74
climate change, 210, 220
Clinton, Bill, 27
Cocconi, Guiseppe, 192–93, 219
comets, 67–68, 105, 164–65, 206
communication: alien messages, 205, 209–16; astrobiology and, 190–95, 199, 205, 210–11, 216–19, 222, 224; Australia and, 204; Bell and, 201–2; Cocconi-Morrison paper and, 192–94; dolphins and, 190, 195–96, 211–20, 224; Drake equation and, 30–33, 190–92, 196, 200, 216–21; electromagnetism and, 192–93, 200, 203–4; Green Bank Observatory and, 192–98, 217, 219; hydrophones and, 213; impostor syndrome and, 217; intelligent species and, 192–93; Legacy Survey of Space and Time (LSST), 203–5; LGM-1 and, 202; Mars and, 29, 206; mathematical logic and, 210; Milky Way and, 193, 219–20; Mullard Radio Astronomy Observatory and, 201–2; 'Oumuamua and, 207–9; Project Ozma, 192–98; SETI, 191, 194–201, 204; spectra and, 200, 203–4; Square Kilometer Array and, 203–5; technosignatures and, 5, 205–7; telescopes and, 192–98, 203–7, 217–19
continental drift, 40
continental shelf, 72
Copernicus, Nicolaus, 10, 13
copper, 50, 72, 169
COVID-19 vaccines, 210
cratons, 84, 87, 90
crystals: AGB stars and, 154; calcium carbonate, 153; Enceladus and, 38; ice, 38, 65; magnetite, 26; meteorites and, 26–27; mineral, 26–27, 38, 60, 65, 153–54, 160, 175–76; sulfur, 60; Widmanstätten pattern, 160; zircon, 175–76
Curiosity rover, 94
cyanobacteria, 82, 89–90

Darwin, Charles, 12–13, 93
Darwin's Lost Worlds (Brasier), 93–94
Deep Space Network (DSN), 63, 100–101
Descartes, René, 19
dinosaurs, 79, 81
Discovery Channel, 161
Discovery Trails to Early Earth (van Kranendonk and Johnston), 76
DNA, 2, 14, 22, 191
dolerite, 77
dolphins: communication and, 190, 195–96, 211–20, 224; hydrophones and, 213; interaction with, 33; Order of the Dolphin, 8, 196, 211, 217, 219–20; *Stenella* and, 212; unique markings, 213; vocalizations of, 13–16; water and, 8, 212–14, 224; Wild Dolphin Project, 212
Doppler, Christian, 120

Doppler effect, 120–22, 125–26, 129, 142
Dragonfly mission, 9, 107
Drake, Frank: Cocconi-Morrison paper and, 192–94; Green Bank conference and, 194; hydrogen and, 193, 197; Mullard Radio Astronomy Observatory and, 201; Project Ozma and, 192–98; radio astronomy and, 29–30, 191–93, 196–98; stars and, 30–31, 131, 149, 192, 196–98, 219, 222
Drake equation: astrobiology and, 29–34; communication and, 30–33, 190–92, 196, 200, 216–21; exoplanets and, 131, 149; Mars and, 216; Order of the Dolphin and, 196, 217, 219–20
drilling core samples, 96, 103–4
DSN (Deep Space Network), 63, 100–101
dust particles: AGB stars and, 155; asteroids and, 167; HL Tau and, 183–86, *185*; Mars and, 105; McKay and, 25; meteorites and, 155–60, 173–74, 183–86, 189; protoplanetary disks and, 155, 173, 183–86, 206

"Early Archean (3.3-Billion to 3.5-Billion-Year-Old) Microfossils from Warrawoona Group, Australia" (Schopf), 90
Earth: Abyssal Zone, 55; age of, 12; Anders's photograph of, 221; Antarctica, 161; Archean eon, 87, 90–92, 94, 108; astronomical units, 122; atmosphere of, 9, 16–18, 55, 82, 85, 135, 141–44, 147–48, 158, 170, 177, 180, 182; continental drift of, 40; Copernicus on, 10, 13; cratons and, 84, 87, 90; debate over dating samples of, 90–94; early conditions of, 175–81; *Galileo* flyby, 16–19; Hadean eon, 55, 87, 176, 180–81, 187–88; magnetism and, 41–42, 63, 85, 150, 193; meteorites and, 158 (*see also* meteorites); as pale blue dot, 1–4, 16, 34, 112, 147, 158, 184, 220–21; Precambrian eon, 88, 108; radioactive core of, 40; rocks and, 6–7, 9, 41–44, 75, 78–79, 84, 87–90, 93–96, 104–9, 114, 141, 148, 153, 156, 158, 161, 172–77, 180, 182, 187, 216, 218; similarity to Mars, 6, 83–84; temperature and, 135; water and, 15, 38, 42–44, 55–56, 63–72, 82, 85–86, 156, 165–66, 173–80, 187, 205, 222, 224
electricity: chlorophyll and, 56; Europa and, 38, 69; oceans and, 56; photons and, 111; photosynthesis and, 55–56; radio isotopes and, 69; renewable, 73; telescopes and, 111
electromagnetism: communication and, 192–93, 200, 203–4; ether and, 192; Europa and, 38; radio and, 17; TV and, 17
ELT. *See* Extremely Large Telescope (ELT)
Enceladus: *Cassini* and, 15, 38–39; crystals and, 38; *Dragonfly* and, 107; geysers of, 15, 59, 65, 67; ice and, 5, 15, 38–39, 56–57, 65, 71; *Orpheus* and, 71; return to, 107; rocks and, 38, 107; search for life and, 5, 8, 37; size of, 38

INDEX [239]

Endeavour (hydrothermal vent system): *Alvin* and, 50–51; hydrothermal vents and, 39, 46–53, 52, 59, 73–74; intense activity of, 46; Juan de Fuca Ridge and, 46, 49; *Nautilus* and, 39; temperature of, 50
Epsilon Eridani, 192, 197
ESO (European Southern Observatory), 139
ether, 192
Eucrite meteorite, 164
Europa: *Dragonfly* and, 107; electricity and, 38, 69; electromagnetism and, 38; *Galileo* and, 37–39, 63, 66; gravity and, 38, 44; ice and, 5, 15, 38–39, 57, 63–71, 74, 136, 218; isotopes and, 69; landing on, 19, 216, 218; magnetism and, 38, 63; oceans and, 37–39, 43–44, 57–59, 63–71, 74, 217; REASON and, 66; rocks and, 38, 43–44, 65–68, 107, 216, 218; ROVs and, 19, 70; water and, 10, 15, 38, 43–44, 59, 63–69
Europa Clipper, 65–68, 218
European Southern Observatory (ESO), 139
European Space Agency, 68, 106
"Evidence from Detrital Zircons for the Existence of Continental Crust and Oceans on the Earth 4.4 Gyr Ago" (*Nature*), 176
Ewing, Maurice, 40–41
exoplanets: astrobiology and, 112, 115, 128, 131–33, 136, 143, 146, 148; astrophysics and, 118–21; atmospheres and, 3, 110, 125, 129, 135–44, 147–51, 218; biomarkers and, 137–38, 141–42, 149–50; Bruno on, 11, 115, 149; Campbell and, 125–26; carbon and, 147; chemistry and, 119, 137–38, 143–44; Chile and, 110, 113–15, 139–40, 142; Doppler effect and, 120–22, 125–26, 129, 142; Drake equation and, 131, 149; 51 Pegasi b, 129–30, 132, 149–50; Frail and, 128; "Goldilocks zone" and, 113, 128, 130, 133, 136, 148; gravity and, 116–18, 121–23, 128; Green Bank conference and, 194; habitability and, 113, 131–52; Heintz and, 123–24; ice and, 135–36; indirect evidence of, 116–18, 121–22; K2-18b, 147; KELT 9b, 130; Kepler 16b, 130; Kepler 37b, 130; Kepler Observatory and, 112–13; metabolism and, 137; methane and, 137, 142, 147, 151; METIS and, 141–42; Milky Way and, 116, 120, 145–47; Newton's laws and, 117; oxygen and, 137, 142, 151; photons and, 6, 111, 120, 124, 140, 142, 148, 151; possibility of, 11; protoplanetary disks and, 133, 136, 147; pulsars and, 127–28; radial velocity method and, 121, 123–25, 130, 146, 183; rocks and, 113–14, 132, 141, 143, 148, 150, 153, 155–63, 171–82, 186–87; spectra and, 114, 118–21, 125, 132, 137–38, 141–43; stars and, 110–32, 140–46, 149–52; Strand and, 123; Struve and, 121–24, 129, 137, 192, 194; telescopes and, 110–11, 113–16, 119–21, 124, 127, 132–33, 138–43, 146–51; tolerable zone and, 132; Transiting Exoplanet

exoplanets (cont.)
Survey Satellite (TESS), 144–46, 206; TRAPPIST and, 132–36, 143, 148–49; Walker and, 125–26; water and, 135–36, 151–52; Wolszczan and, 128; Yang and, 125–26
Expedition 2017—Wiring the Abyss, 45
Extremely Large Telescope (ELT): abilities of, 140–42, 148–49, 218; ALMA and, 183; Chile and, 139–42, 148–49, 183, 204, 218; description of, 139; exoplanets and, 139–42, 148–49; James Webb Space Telescope and, 148

fiber-optic cable, 45–46, 49, 100
51 Pegasi b, 129–30, 132, 149–50
first light, 114, 140, 203
first-rank stars, 110–11
formaldehyde, 178
fossils: Apex, 90–92; Archean eon and, 87, 90–92, 94, 108; asteroids and, 79; astrobiology and, 83, 91, 98; Australia and, 6, 25, 32, 75–77, 81, 84, 87–93, 108, 216; bacteria and, 75, 80–82, 88–90; Brasier and, 91–94; carbon and, 80, 82–83, 91–92; chemistry and, 82–83, 108; cratons and, 84, 87, 90; debate over dating of, 90–94; dinosaur, 79, 81; geology and, 77, 79, 84–90, 98; inclusions and, 77; Mars and, 75, 83–86, 94–108; microbes and, 82–83, 91; minerals and, 2, 26, 77, 83, 88, 91–92, 98–99, 103, 108; North Pole Dome and, 90; Precambrian eon and, 88, 108; radioactive dating of, 79, 98; rocks and, 2, 6–7, 75, 77–80, 84–99, 103–9; samples and, 80, 88, 92–99, 104–8; Schopf on, 90–94; sedimentary rock and, 75, 79, 88–89, 108; Strelley Pool and, 88, 90; stromatolites and, 75–76, 81–83, 87–88, 109, 224; Tyler-Barghoorn paper and, 89–90; volcanoes and, 79, 85, 98
Frail, Dale, 128
Fraunhofer, Joseph von, 118, *119*
frost line, 173–74, 177
fusion crusts, 24, 158, 161, 163–64

Gaia, 18, 144–46
Galápagos rift, 42, 54
Galileo space probe: astrobiology and, 15–19, 39, 66; *Atlantis* launch of, 16; atmospheres and, 16–18; Earth flyby of, 16–19; Europa and, 37–39, 63, 66; gravitational assist maneuvers of, 16; Jupiter and, 37–39, 63, 66; NASA and, 37; pale blue dot and, 16; Sagan and, 16–17; science team of, 38
Galle, Johann, 117–18
Gamma Cephei, 126, 129
geology: bacteria and, 37, 89; fossils and, 77, 79, 84–90, 98; Green Bank conference and, 195; heat and, 37, 40, 42, 54, 89, 171; Mars and, 25, 84–86, 98; McKay and, 25–28; meteorites and, 171, 175–76; oceans and, 195; rocks and, 2 (*see also* rocks)
geysers, 15, 59, 65, 67
"Goldilocks zone," 113, 128, 130, 133, 136, 148
GPS, 76–78, 81, 109

gravity: asteroids and, 105; Churyumov-Gerasimenko and, 68; Copernicus on, 10, 13; Europa and, 38, 44; exoplanets and, 116-18, 121-23, 128; *Hayabusa2* and, 105; HL Tau and, 184; Jupiter and, 117; Mars and, 105; meteorites and, 160, 166-68; Newton and, 117-18; oceans and, 37, 39-40, 42, 45-46, 53-54; radial velocity method and, 121, 123-25, 130, 146, 183; Strand and, 123; sun and, 116-17, 173
Green Bank Observatory: Cocconi-Morrison paper and, 192-94; communication and, 192-98, 217, 219; Epsilon Eridani and, 192, 197; Project Ozma and, 192-98; Tau Ceti and, 192, 197
Gujba meteorite, 160
Gunflint cherts, 88-89

habitability, 7; atmospheric conditions and, 17-18, 85-86, 104, 135-38, 141, 143-44, 147-50; exoplanets and, 131-52; "Goldilocks zone" of, 113, 128, 130, 133, 136, 148; Mars and, 86-87; METIS and, 141-42; tolerable zone and, 132; water and, 85, 135-36
Hadean eon: fossils and, 87; meteorites and, 176, 180-81, 187-88; oceans and, 55
Hale, George Ellery, 140
Hale telescope, 121
Haleakala volcano, 207
Hanson, Robin, 221
Harvard University, 120, 193
Hawaii, 140, 142-43, 206-7

Hayabusa1 probe, 165
Hayabusa2 probe: asteroids and, 105, 165-72, 209; comets and, 165; gravity and, 105; meteorites and, 165, 170-72; reentry probe of, 170; Ryugu and, 165-71, *167*; Small Carry-On Impactor and, 169
Heezen, Bruce, 40-41
Heintz, Wulff, 123-24
helium, 154
Hercules (ROV): hydrothermal vents and, 48, 51; *Nautilus* and, 36, 46, 49, 100; oceans and, 36, 48-51, 53, 74, 100
HIBOU minilander, 166
Hitchhiker's Guide to the Galaxy, The (Adams), 190
HL Tau, 183-86, *185*
Holmes, Arthur, 40
Homer, 72, 189
Homo sapiens, 112, 211, 215
Hooker telescope, 140
Howardite meteorite, 164
HR 8974 (Gamma Cephei), 126, 129
Huang, Su-Shu, 194
Hubble, Edwin, 12-13
Hubble Space Telescope, 148
hydrogen: abundance of, 193; astrobiology and, 14; deuterium, 174; Drake and, 193, 197; meteorites and, 154, 173-74; oceans and, 56-57, 60; radio astronomy and, 193, 197; spectra of, 193
hydrogen cyanide, 173
hydrophones, 213
hydrothermal vents: *Alvin* and, 58; *Argus* and, 48; astrobiology and, 5, 15, 29, 32; Ballard and, 45; chemosynthesis and,

INDEX

hydrothermal vents (cont.)
57–61, 64; chimneys and, 37,
52–54; clams and, 42, 57; depth
of, 5; discovery of, 42–43,
54; Endeavour, 39, 46–53, 59,
73–74; Galápagos rift and, 54;
geological puzzle of, 42; *Hercules* and, 48, 51; Juan de Fuca
Ridge and, 46, 49; mid-ocean
ridges and, 43; *Nautilus* and, 39,
45; oceans and, 5, 15, 29, 32, 39,
42–64, 71–73, 80; Pacific Ocean
and, 59; pressure at, 55; shrimp
and, 53–54, 60–62; Solwara 1,
73; subsurface plumbing of,
53; tube worms and, 53–54, 57,
59–60, 224

ice: Antarctica and, 161; crystals
of, 38, 65; Enceladus and, 5, 15,
38–39, 56–57, 65, 71; Europa
and, 5, 15, 38, 57, 63–71, 74,
136, 218; Europe and, 65–70;
exoplanets and, 135–36; frost
line and, 174; Mars and, 66, 86,
218; meteorites and, 35, 174,
184; oceans and, 39
impostor syndrome, 217
inclusions: fossils and, 77;
meteorites and, 156–57, 170–71,
181, 186
infrared photons, 61–62, 135, 141
Ingenuity drone, 102, 107
Institute of Astronomy, 113
iron: Australia and, 75–76, 80, 88;
Campo de Cielo and, 159–60;
meteorites and, 159–63; mining
of, 75–76; oceans and, 50, 53,
72, 80
isotopes: chemistry of, 108; Europa
and, 69; meteorites and, 156–58,

171, 174–76, 187; radioactive,
25, 69

James Webb Space Telescope, 107,
146–48, 150
Jannasch, Holger, 58, 63
Japan, 8, 105, 165–66, 170
JAXA, 165–66, 170
Jet Propulsion Laboratory (JPL), 67,
70, 99–102
Jezero crater, 98, 104
Johnston, Jean, 76
Journey to the Center of the Earth
(Verne), 70
JPL (Jet Propulsion Laboratory), 67,
70, 99–102
Juan de Fuca Ridge, 46, 49
Jupiter: astrobiology and, 2, 5, 16,
32, 34; Deep Space Network
and, 63; dominance of, 116;
equations for, 116–17; Europa
and, 37–39 (*see also* Europa);
Europa Clipper and, 65–68,
218; frost line and, 174; *Galileo*
and, 37–39, 63, 66; as giant gas
planet, 5; gravity and, 116–17;
magnetism and, 38; moons of, 2,
5, 32, 34, 37–38, 63, 71, 130; orbit
of, 116; similar discoveries like,
118, 122–23, 126, 129–30, 146;
velocity imparted to Sun by,
117

K2-18b, 147
Keck telescopes, 140, 206
KELT 9b, 130
Kepler, Johannes, 145
Kepler 16b, 130
Kepler 37b, 130
Kepler observatory, 112–13
KIC 8462852 (Tabby's star), 205–6

INDEX [243]

Kirchhoff, Gustav, 119
Knorr (research vessel), 42, 45, 54

La Silla Observatory, 110, 114, 132–33, 138, 150
Lake Superior, 88
lava, 42, 66–67
Le Verrier, Urbain, 117–18
Lederberg, Joshua, 194
Legacy Survey of Space and Time (LSST), 203–5
Levi, Primo, 153–54
Leviathan of Parsonstown, 139–40
LGM-1 (Little Green Men), 202
Lick Observatory, 129
LIDAR (light detection and ranging), 68
Life on Earth (TV series), 211
light detection and ranging (LIDAR), 68
lightning, 17
Lilly, John, 195–96
limestone, 77, 79–80, 153
Llano de Chajnantor, 181–82, 187
Loeb, Avi, 208
Lovelock, James, 18, 222
Lowell, Percival, 22–23, 206, 208
LSST (Legacy Survey of Space and Time), 203–5
Lyndon B. Johnson Space Center, 24–25

magnetism: communication and, 193, 200, 203–4; Earth and, 41–42, 63, 85, 150, 193; ether and, 192; Europa and, 38, 63; field reversal and, 41–42; Jupiter and, 38; meteorites and, 162; minerals and, 26–27; M-type stars and, 150; radio, 17; rocks and, 41, 162; TV, 17; volcanoes and, 85

magnetite, 26–27
magnetometers, 38, 63
Man and Dolphin (Lilly), 195–96
Marcy, Geoff, 129
Mars: ALH84001 and, 26, 94; astrobiology and, 3–8, 19, 22–29, 32, 34, 83, 98, 218; atmosphere of, 67, 85–86, 104, 218; canals of, 22–23, 206, 208; communication and, 29, 206; cratons and, 84, 87, 90; *Curiosity* rover and, 27–28, 94; Drake equation and, 216; fossils and, 75, 83–86, 94–108; frost line and, 174; geology and, 25, 84–86, 98; gravity and, 105; habitability of, 86–87; ice and, 66, 86, 218; Jezero crater, 98, 104; Lowell and, 22–23, 206, 208; minerals and, 108; NASA and, 23–28, 67–68, 94–98, 105–7; Noachian period of, 25; oceans and, 25, 58, 66–68; old surface rocks of, 84; *Opportunity* rover and, 94; *Perseverance* rover and, 68, 94–108; *Phoenix* lander and, 24; return project of, 106–7; rocks and, 2–3, 6, 25, 67–68, 75, 84–85, 93–99, 104–9, 174, 216, 218; ROVs and, 19, 32, 85, 95–96, 104, 106, 109; similarity to Earth, 6, 83–84; *Sojourner* rover and, 94; spectra and, 23, 82; *Spirit* rover and, 94; stromatolites and, 75, 83; *Viking* landers and, 23–24, 28, 58, 67–68; volcanoes and, 85–86; water and, 25, 85–86, 98, 174
Mars Reconnaissance Orbiter (MRO), 97, 101
mathematics: Doppler and, 120; Drake equation, 190 (*see also* Drake equation); gravity, 44,

mathematics (cont.)
116–17; habitability and, 134–35;
universal logic of, 210
Maury, Antonia, 120–21
Mayor, Michel, 129
McKay, David, 25–28
Mercury, 13, 114, 128–29, 134
metabolism: astrobiology and, 3, 17, 20–23, 27; biomarkers and, 137; Brasier paper on, 92; carbon cycle and, 195; exoplanets and, 137; oceans and, 55–56, 58, 60, 64, 67; photosynthesis and, 82
Meteorite Men (TV series), 161
meteorites: ALH84001, 24–28, *26*, 94; Allende, 156–57, *157*; asteroids and, 7–8, 32, 155–60, 164–74, 177–81, 189; astrobiology and, 176–79, 189; Australia and, 156; best time to look for, 161–62; calcium-aluminum inclusions and, 156, 186–87; cameras and, 164, 166; Campo de Cielo, 159–60; carbon and, 25–26, 153–60, 165–66, 170–74, 177–81, 186; chemistry and, 153–57, 164–65, 168–81, 188; Chile and, 173, 181; chondrite and, 156–59, 165, 170, 174, 177–80, 186; crystals and, 26–27; dust particles and, 155–60, 173–74, 183–86, 189; Earth's atmosphere and, 159, 164, 170; Eucrite, 164; frost line and, 174; fusion crusts and, 24, 158, 161, 163–64; geology and, 171, 175–76; gravity and, 160, 166–68; Gujba, 160; *Hayabusa2* and, 165, 170–72; Howardite, 164; hydrogen and, 154, 173–74; ice and, 35, 174, 184; inclusions and, 156–57, 170–71, 181, 186; as inert fragments, 178; iron and, 159–63; isotopes and, 156–58, 171, 174–76, 187; Llano de Chajnantor and, 181–82, 187; magnetism and, 162; and meteowrongs, 88, 162; minerals and, 153–54, 156, 160, 164, 171, 175–78; Murchison, 156, 178; NASA and, 24–25, 165; *New Horizons* and, 155; NWA (North West Africa), 159–64; oxygen and, 154, 175–76; protoplanetary disks and, 155, 173, 181, 184, 186–87; radioactivity and, 156, 171, 176, 187; regmaglypts and, 159, 164; rocks and, 7, 24–25, 88, 94, 156–63, 171, 174–80, 186; samples of, 8, 25, 32, 94, 160, 164–72, 178–79, 189; sun and, 154–55, 171, 173; surfaces of, 158–59; US Antarctic Search for Meteorites, 24; *Voyager 2* and, 155; water and, 156, 165–66, 171–72; where to look for, 161; Widmanstätten pattern and, 160
methane: astrobiology and, 16–18, 80, 137, 142, 147, 151; atmospheric, 16–18, 80, 137, 142, 147, 151; exoplanets and, 137, 142, 147, 151; spectra of, 16
METIS (Mid-infrared ELT Imager and Spectrograph), 141–42
microbes: astrobiology and, 17–18, 23, 26–27, 29; fossils and, 82–83, 91; oceans and, 55–62; omnivorous, 82–83; oxygen and, 14, 57, 64, 82–83; stromatolites and, 82–83

INDEX [245]

microcosmos, 10, 13–15
"Microfossils of Sulphur-Metabolizing Cells in 3.4-Billion-Year-Old Rocks of Western Australia" (Brasier et al.), 92
Mid-infrared ELT Imager and Spectrograph (METIS), 141–42
mid-ocean ridges, 41–43, 49, 54, 84
Milky Way: astrobiology and, 12, 30, 32–33; communication and, 193, 219–20; exoplanets and, 116, 120, 145–47; hydrogen spectra and, 193
Miller, Stanley, 180
minerals: astrobiology and, 2, 26–28; bacteria and, 27; chimneys and, 37, 52–54; crystals, 26–27, 38, 60, 65, 153–54, 160, 175–76; fossils and, 2, 26, 77, 83, 88, 91–92, 98–99, 103, 108; Mars and, 108; meteorites and, 153–54, 156, 160, 164, 171, 175–78; oceans and, 37, 50–51, 57, 60, 72–73; towers of, 37, 51, 53; water and, 50, 98, 176–77. See also specific minerals
mining, 73, 75–76
Mizar A, 120
Mizar B, 120
Morocco, 161–64
Morrison, Philip, 192–94, 219
Mount Wilson, 140
MRO (Mars Reconnaissance Orbiter), 97, 101
Mullard Radio Astronomy Observatory, 201–2
Murchison meteorite, 156, 178
"My Oldest Fossils Are Older Than Your Oldest Fossils" (MOFAOTYOF), 93

NASA: *Apollo* program, 25, 141, 166, 221; Astrobiology Institute, 28; budget, 107; *Cassini*, 15, 38–39; *Curiosity*, 94; Deep Space Network (DSN), 63, 100; *Dragonfly*, 9, 107; *Europa Clipper*, 65–68, 218; European Space Agency (ESA) and, 106; *Galileo*, 37 (see also *Galileo* space probe); Hubble Space Telescope, 148; JAXA and, 165; Jet Propulsion Laboratory (JPL), 67, 70, 99–102; *Kepler*, 111–12; Lyndon B. Johnson Space Center, 24–25; Mars and, 23–26, 28, 67–68, 94, 97–98, 105–7; McKay and, 25, 27–28; meteorites and, 24–25, 165; off-world autonomous driving and, 102; *Opportunity*, 94; *Orpheus*, 70–71; OSIRIS-REx, 105, 165, 172, 178, 209, 218; *Perseverance*, 68, 94–108; *Phoenix*, 24; rocks and, 25, 38, 68, 98, 105–7, 172; SETI, 194; *Sojourner*, 94; *Spirit*, 94; *Stardust*, 165; Transiting Exoplanet Survey Satellite (TESS), 144–46, 206; *Voyager 1*, 1, 15, 34; *Voyager 2*, 63, 155
National Academy of Sciences, 194
Nature (journal), 91, 129, 176, 192
Nautilus (exploration vessel): *Argus* and, 36, 46, 49, 100; conditions aboard, 47–48; *Endeavour* and, 39; *Hercules* and, 36, 46, 49, 100; hydrothermal vents and, 39, 45; Ocean Exploration Trust and, 45; oceans and, 36, 39, 45–49, 100, 115; Pacific Ocean and, 36, 45

nebulae, 128, 157, 173, 178
Nelson, Bill, 107
Neptune, 117, 147–48
NEPTUNE observatory, 45–46
neutron stars, 127, 202
New Horizons spacecraft, 155
New Technology Telescope (NTT), 114–15
Newton, Isaac, 117–18
nickel, 160
Nigeria, 160
nitrogen, 3, 14, 80, 92
Noachian period, 25
Nobel Prize, 28, 126, 180, 195–96, 201
North Pole Dome, 90
North West Africa (NWA) meteorites, 159–64
Northern Territory, 175
NTT (New Technology Telescope), 114–15
NWA 6135, 159

"Occurrence of Structurally Preserved Plants in Pre-Cambrian Rocks of the Canadian Shield" (Tyler and Barghoorn), 89
Ocean Exploration Trust, 45
Ocean Networks Canada (ONC), 45
oceans: Abyssal Zone, 55; ANGUS and, 42; *Argus* and, 36, 45, 48–49, 74, 100; astrobiology and, 39, 45, 57–59, 63–64, 66, 71, 74; bacteria and, 37, 54, 60–64; benthic and resistivity sensors (BARS) and, 51, *52*; carbon and, 55–60, 73; Challenger Deep, 44; chemistry and, 37, 43, 50–51, 54–58, 62, 64, 66; chimneys in, 37, 52–54; clams and, 42, 57; continental shelf, 72; demographics of, 43–44; DNA and, 2; early mapping of, 40–41; electricity and, 56; Enceladus and, 15, 38–39; Endeavour and, 39, 46–53, 59, 73–74; Europa and, 37–39, 43–44, 57–59, 63–71, 74, 217; geology and, 37, 39–40, 42, 45–46, 53–54, 195; Heezen-Tharp map, 40–41; Hercules and, 36, 48–51, 53, 74, 100; hydrogen and, 56–57, 60; hydrothermal vents and, 5, 15, 29, 32, 39, 42–64, 71–73, 80; ice and, 39; iron and, 50, 53, 72, 80; Juan de Fuca Ridge, 46, 49; *Knorr* and, 42, 45, 54; lack of knowledge of, 39–40; Mars and, 25, 58, 66–68; metabolism and, 55–56, 58, 60, 64, 67; microbes and, 55–62; mid-ocean ridges, 41–43, 49, 54, 84; minerals and, 37, 50–51, 57, 60, 72–73; *Nautilus* and, 36, 39, 45–49, 100, 115; *Orpheus* and, 70–71; oxygen and, 56–57, 60, 64–65; rift zones, 42, 54, 66; ROVs and, 32; samples from, 51, 53, 58–59, 66; seafloor pressure, 55; surface, 39, 44, 65, 85; tectonic plates and, 42–43, 46, 49–50, 54; temperature and, 42, 50–51, 54–55, 61; towers in, 37, 51, 53; *Viking* landers and, 58, 67–68; volcanoes and, 51, 67; Woods Hole Oceanographic Institution, 70–71
Odyssey, The (Homer), 72
Oliver, Barney, 194
On the Infinite Universe and Worlds (Bruno), 11
On the Revolutions of the Celestial Spheres (Copernicus), 10, 13

ONC (Ocean Networks Canada), 45
Opportunity rover, 94
Opticks (Newton), 118
Order of the Dolphin, 8, 196, 211, 217, 219–20
Orpheus (AUV), 70–71
OSIRIS-REx mission, 105, 165, 172, 178, 209, 218
'Oumuamua, 207–9
Outermost House, The (Beston), 223–24, *223*
OWL minilander, 166
oxygen: Archean, 87; astrobiology and, 3, 14, 16–17; atmospheric, 16–17, 80, 82, 87, 137, 142, 151; bacteria and, 60, 64, 82; cyanobacteria and, 83; exoplanets and, 137, 142, 151; meteorites and, 154, 175–76; microbes and, 14, 57, 64, 82–83; oceans and, 56–57, 60, 64–65; photosynthesis and, 55–57, 62, 64, 82, 89, 153, 195; spectra of, 16

Pacific Ocean, 110, 216; hydrothermal vents and, 5, 45–46, 49, 59; Juan de Fuca Ridge, 46, 49; *Nautilus* and, 36, 45; sea floor spreading and, 58; tectonic plates and, 46, 49, 54
Packer, Bonnie, 90
pale blue dot: Anders's photograph and, 220; exoplanets and, 147; *Galileo* and, 16; HL Tau and, 184; as home, 1–2; impact of image, 1–2; Sagan and, 1–4, 220; vantage point of, 112, 158; *Voyager 1* and, 1–4
Palomar Observatory, 121
Pan-STARRS telescope, 207

Papua New Guinea, 73
Parsons, William, 139–40
Pearman, Peter, 194
Periodic Table, The (Levi), 153–54
Perseverance rover: cameras of, 99, 101–3; cosmic radiation and, 99; cost of, 102; Deep Space Network (DSN) and, 100–101; *Ingenuity* drone and, 102, 107; Jet Propulsion Laboratory (JPL) and, 99–101; Jezero crater and, 98, 104; long exploration of, 102, 104–5; Mars and, 68, 94–108; off-world autonomous driving and, 102; rocks and, 68, 96, 99, 103–8; science team of, 99–100, 103, 107–8; site-selection for, 97–98
Philae lander, 68
Phoenix lander, 24
photoautotrophs, 56
photography: Anders and, 220; Daguerreotypes, 159; exoplanets and, 119–20, 124; fossils and, 88–90, 109; Gunflint cherts and, 88–89; spectra and, 119–20
photons: exoplanets and, 6, 111, 120, 124, 140, 142, 148, 151; HL Tau and, 186; infrared, 61–62, 135, 141; photosynthesis and, 62, 66; quantum flips and, 193; solar, 55–56, 61; telescopes and, 111, 120, 124, 140, 142, 148; thermal, 61; ultraviolet, 168
photosynthesis: astrobiology and, 3; carbon and, 55–57, 62, 64, 82, 89, 153, 195; chlorophyll and, 56, 82; electricity and, 55–56; metabolism and, 82; oxygen and, 55–57, 62, 64, 82, 89, 153, 195; photons and, 62, 66; plants

photosynthesis (cont.)
    and, 55–57, 62, 64, 82, 89, 153, 195
Pickering, Edward, 120–21
Pilbara Craton, 87, 90
Pilbara Shire, 90
Pluto, 185
Precambrian eon, 88, 108
preponderance of evidence, 92
Project Ozma: Atchley and, 194; Bassham and, 195; Benson and, 195; Breakthrough Listen and, 198; Calvin and, 195–96; Drake and, 192–98; Green Bank Observatory and, 192–98; Lilly and, 195–96; Oliver and, 194; SETI and, 196–98
proof of life, 27–28, 94
"Proposal for a Project of High-Precision Stellar Radial Velocity Work" (Struve), 121
Proterozoic eon, 108
protoplanetary disks: ALMA and, 183–88; carbon and, 155, 173; dust particles and, 155, 173, 183–86, 206; exoplanets and, 133, 136, 147; HL Tau and, 183–86, *185*; meteorites and, 155, 173, 181, 184, 186–87; Tabby's star and, 206
Proxima Centauri, 62
PSR 1257+12, 127–28, 130
pulsars, 127–28, 202
pyroclastic rock. *See* inclusions

quantum mechanics, 193
Queloz, Didier, 129
"Questioning the Evidence for Earth's Oldest Fossils" (Brasier), 91

Radar for Europa Assessment and Sounding (REASON), 66
radial velocity method: exoplanets and, 121, 123–25, 130, 146, 183; spectra and, 121, 125; Struve and, 121, 124
radiation: Europa and, 66; hydrothermal vents and, 61–62; infrared, 61–62, 135; Mars and, 99; ultraviolet, 168–69
radio astronomy: Drake and, 29, 191–93, 196–98; Mullard Radio Astronomy Observatory and, 201–2; Project Ozma and, 192–98
radioactivity: carbon dioxide and, 58; dating methods and, 79, 98; Earth's core and, 40; fossils and, 79; meteorites and, 156, 171, 176, 187
REASON (Radar for Europa Assessment and Sounding), 66
regmaglypts, 159, 164
remotely operated vehicles (ROVs): advantages/disadvantages of, 95–96; dust particles and, 105; Europa and, 19, 70; Mars and, 19, 32, 85, 95–96, 104, 106, 109; oceans and, 32, 36, 46–49, 100; samples and, 6 (*see also* samples); search for life and, 5–6, 19; site-selection for, 97–98. *See also specific rovers*
Ridgeia piscesae (tube worm), 59–60, 52
rift zones, 42, 54, 66
Riftia pachyptila (tube worm), 59–60
Rimicaris exoculata (shrimp), 60–62
RNA, 59
rocks: Australia and, 6, 25, 75, 77, 84, 87–93, 108, 156, 216;

INDEX [249]

composition of, 20, 99, 141; Earth and, 6–7, 9, 41–44, 75, 78–79, 84, 87–90, 93–96, 104–9, 114, 141, 148, 153, 156, 158, 161, 172–77, 180, 182, 187, 216, 218; Enceladus and, 38, 107; Europa and, 38, 43–44, 65–68, 107, 216, 218; exoplanets and, 113–14, 132, 141, 143, 148, 150, 153, 155–63, 171–82, 186–87; fossils and, 2, 6–7, 75, 77–80, 84–99, 103–9; high cost of returning, 96; magnetism and, 41, 162; Mars and, 2–3, 6, 25, 67–68, 75, 84–85, 93–99, 104–9, 174, 216, 218; meteorites, 7, 24–25, 88, 94, 156–63, 171, 174–80, 186; NASA and, 25, 38, 68, 98, 105–7, 172; oceans and, 41–44, 50–53; *Perseverance* rover and, 68, 96, 99, 103–8; samples of, 2, 6, 20, 25, 51, 53, 80, 88, 92–99, 104–8, 160, 171–72, 178, 218; sedimentary, 75, 79, 88–89, 108

ROVs. *See* remotely operated vehicles (ROVs)

Ryugu: asteroids and, 165–71, 179, 189; chondrite and, 170–71; *Hayabusa2* and, 165–71, *167*; sample materials from, 170–72; structure of, 171

Sagan, Carl: Anders's photograph of Earth and, 221; cosmic context and, 221; Drake and, 194; *Galileo* and, 16–17; Green Bank conference and, 194; Lederburg and, 194; life criteria of, 17, 23; pale blue dot and, 1–4, 220; *Voyager 1* and, 1

Sahara desert, 8, 153, 161, 163

samples: asteroids and, 8, 32, 105, 164–72, 178, 189, 209, 218; astrobiology and, 19–25, 32; Doppler effect and, 125; fossils and, 80, 88, 92–99, 104–8; meteorites and, 8, 25, 32, 94, 160, 164–72, 178–79, 189; ocean, 2, 51, 53, 58–59, 66; rock, 2, 6, 20, 25, 51, 53, 80, 88, 92–99, 104–8, 160, 171–72, 178, 218; soil, 20, 23, 58

sandstone, 79

satellites, 38, 48, 84, 98, 144

Saturn: astrobiology and, 2, 5, 9, 32, 34; *Cassini* and, 38; *Dragonfly* and, 9, 107; Enceladus, 5, 8, 15, 37–39, 57, 59, 65, 67, 71, 107; as giant gas planet, 5; moons of, 2, 5, 9, 32, 34, 37–38, 71, 107, 184; rings of, 184; Titan, 9, 67, 107

Schopf, William, 90–94

Science (journal), 27, 89

science teams: *Galileo*, 38; *Perseverance*, 99–100, 103, 107–8; *Viking*, 24

Scientific American (journal), 208

Score, Roberta, 24

search for extraterrestrial intelligence. *See* SETI

"Search for Substellar Companions to Solar-Type Stars, A" (Walker, Campbell, and Yang), 126

secondary processing, 89

sedimentary rock, 75, 79, 88–89, 108

SETI: Breakthrough Listen and, 198; communication and, 191, 194–201, 204; NASA and, 191, 194–201, 204; Project Ozma and,

SETI (*cont.*)
196–98; stars and, 196–200; US Congress and, 199
shrimp, 53–54, 60–61
silicon carbide, 154–55, 157, 158
61 Cygnus, 123
Small Carry-On Impactor, 169
Sojourner rover, 94
Solwara 1, 73
SpaceX Falcon Heavy rocket, 65
spectra: absorption lines and, 120, 125, 142; Bunsen and, 119; cameras and, 132, 141; chlorophyll and, 17, 82; communication and, 200, 203–4; Doppler effect and, 120–22, 125–26, 129, 142; exoplanets and, 114, 118–21, 125, 132, 137–38, 141–43; Fraunhofer and, 118, *119*; hydrogen, 193; Kirchhoff and, 119; Mars and, 23, 82; Maury and, 120–21; methane, 16; METIS and, 141–42; Newton and, 118; oxygen, 16; photography and, 119–20; Pickering and, 120–21; of stars, 16, 114, 118–21, 125, 138, 141–42, 193, 200, 203; sun and, 118–19, *119*; Wollaston and, 118
spinifex, 78–80
Spirit rover, 94
Sproul Observatory, 123–24
Square Kilometer Array, 203–5
Stardust probe, 165
stars: AGB, 154–55; ALMA and, 173, 183–84, 186; Beta Pictoris, 173; Big Bang and, 188; Big Dipper, 120; binary, 113, 120–23, 125, 142; Bruno and, 115; dead, 158; Doppler effect and, 120–22, 125–26, 129, 142; Drake and, 30–31, 131, 149, 192, 196–98, 219, 222; Epsilon Eridani, 192, 197; exoplanets and, 11, 110–32, 140–46, 149–52; first magnitude, 110–11; Gamma Cephei, 126, 129; HL Tau, 183–86, *185*; hydrogen gas between, 193; M-dwarf, 144–45; Milky Way and, 12 (*see also* Milky Way); Mizar A, 120; Mizar B, 120; neutron, 127, 202; odds of planets orbiting, 131; parent, 183; Proxima Centauri, 62; radial velocity method and, 121, 123–25, 130, 146, 183; red, 3, 114, 143, 147, 157; red dwarf, 62; second magnitude, 111; SETI and, 196–200; 61 Cygnus, 123; spectra of, 16, 114, 118–21, 125, 138, 141–42, 193, 200, 203; supernovas, 128, 189, 202; T Tauri, 184; Tabby's star (KIC 8462852), 205–6; Tau Ceti, 192, 197; telescopes and, 110–11, 113, 120–21, 124, 140, 143, 149–51, 182, 192, 197–98, 203; TRAPPIST-1, 133, *134*; truly dark skies and, 112; young, 173, 207
Stenella research vessel, 212
Strand, Kaj, 123
Strelley Pool Chert, 88, 90
stromatolites: age of, 75; astrobiology interest in, 83; Australia and, 75–76, 81, 88; cyanobacteria and, 83; fossils and, 75–76, 81–83, 87–88, 109, 224; Hamelin Pool, 81; Johnston on, 76; Mars and, 75, 83; microbes and, 82–83; as single-celled organisms, 81–82; Strelley Pool Chert and, 88, 90; van Kranendonk on, 76

Struve, Otto: Cocconi–Morrison paper and, 192; exoplanets and, 121–24, 129, 137, 192, 194; 51 Pegasi b and, 129; Green Bank Observatory and, 192, 194; Palomar Observatory and, 121; "Proposal for a Project of High-Precision Stellar Radial Velocity Work," 121; tight orbits and, 122
sugars, 55, 57, 60, 82
sulfides, 50, 53, 56–57, 60, 64
sulfur, 9, 14, 60, 82–83, 92
sun: Abyssal Zone and, 55; astronomical twilight and, 111; astronomical units and, 122; Bruno on, 11; Copernicus on, 10, 13; gravity and, 116–17, 173; infant, 155; meteorites and, 154–55, 171, 173; Milky Way and, 12; photons and, 55–56, 61; photosphere of, 61; Proxima Centauri and, 62; setting, 110; spectra of, 118–19, *119*; supernovas and, 189
supernovas, 128, 189, 202
Swiss T70 telescope, 114, 132

T Tauri stars, 184
Tabby's star (KIC 8462852), 205–6
Tau Ceti, 29, 192, 197
Taurus constellation, 183
technosignatures, 5, 205–7
tectonic plates, 42–43, 46, 49–50, 54
telescopes: ALMA, 173, 182–87; Arecibo, 127; cameras and, 6, 111, 124, 132, 146, 203; communication and, 192–98, 203–7, 217–19; electricity and, 111; exoplanets and, 110–16, 119–21, 124, 127, 132–33, 138–43, 146–51; Extremely Large Telescope (ELT), 139–42, 148–49, 183, 204, 218; first light of, 114, 140, 203; giant, 7–8, 18–19, 114, 121, 127, 142, 182–83, 204, 206; Green Bank Observatory, 192–98, 217, 219; Hale, 121; Hooker, 140; Hubble Space Telescope, 148; improved, 22–23; James Webb Space Telescope, 107, 146–48, 150; Keck, 140, 206; La Silla Observatory, 110, 114, 132–33, 138, 150; large, 18, 121, 124, 132, 139–43, 146, 148–49, 183, 218; Legacy Survey of Space and Time (LSST) and, 203–4; Llano de Chajnantor, 181–82, 187; New Technology Telescope (NTT), 114–15; Palomar Observatory, 121; Pan-STARRS, 207; photons and, 111, 120, 124, 140, 142, 148; radial velocity method and, 121, 123–25, 130, 146, 183; radio, 8, 33, 127, 181–82, 192–93, 197, 204; Square Kilometer Array, 203–4; stars and, 110–11, 113, 120–21, 124, 140, 143, 149–51, 182, 192, 197–98, 203; Swiss T70, 114, 132; Thirty Meter Telescope (TMT), 142–43; as time machines, 87; TRAPPIST, 132–36, 143, 148–49; Vera C. Rubin Observatory, 203–4
temperature: AGB stars, 154; ANGUS and, 42; atmospheric, 55, 129, 135–36, 177; Australia and, 77; cryostatic, 141; Earth and, 135; frost line and, 173–74; Gamma Cephei, 129; Kepler 37b, 130; METIS and, 141; oceans and, 42, 50–51, 54–55, 61
TESS (Transiting Exoplanet Survey Satellite), 144–46, 206

Tharp, Marie, 40–41
theory of evolution: cell division and, 80; Darwin and, 12–13, 93; dolphins and, 211; metabolism and, 21, 58, 67; Miller–Urey experiment, 180; oxygen-producing microbes and, 82; temperature and, 61–62
Thirty Meter Telescope (TMT), 142–43
Titan, 9, 67, 107
Titanic (ship), 45
TMT (Thirty Meter Telescope), 142–43
tolerable zone, 132
towers, 37, 51, 53
Trans-Antarctic mountain chain, 70
Transiting Exoplanet Survey Satellite (TESS), 144–46, 206
Transiting Planets and Planetesimals Small Telescope (TRAPPIST), 132–36, 143, 148–49
TRAPPIST-1, 133, *134*
trophosome, 60
tube worms, *52*, 53–54, 57, 59–60, 224
Twenty Thousand Leagues Under the Sea (Verne), 46
2010: Odyssey Two (Clarke), 65
Tyler, Stanley, 89–90

ultraviolet photons, 168
Underwater Networked Experiments, 45–46
Upsilon Andromeda, 130
Uranus, 117, 147–48, 150
Urey, Harold, 180
US Antarctic Search for Meteorites, 24

van de Kamp, Peter, 124
van Kranendonk, Martin, 76
van Leeuwenhoek, Antonie, 13–14, 205
Venus, 9, 16, 67, 130, 160
Vera C. Rubin Observatory, 203–4
Verne, Jules, 46, 70
Vesta asteroid, 164
Viking landers, 23–24, 28, 58, 67–68
volcanoes: basalt, 51; fossils and, 79, 85, 98; Haleakala, 207; lava, 42, 66–67; magnetism and, 85; Mars and, 85–86; Maunakea, 142; ocean, 51, 67; radioactive dating and, 79; shield, 108
Voyager 1 spacecraft, 1–4, 14–15, 34
Voyager 2 spacecraft, 63, 155

W. M. Keck Observatory, 140
Walker, Gordon, 125–26
water: ANGUS and, 42; bacteria and, 60, 81–82; benthic and resistivity sensors (BARS) and, 51, *52*; condensation and, 43, 174, 177; dolphins and, 8, 212–14, 224; drilling, 69; Earth and, 15, 38, 42–44, 55–56, 63, 65–72, 82, 85–86, 156, 165–66, 173–74, 176–80, 187, 205, 222, 224; Europa and, 10, 15, 38, 43–44, 59, 63–69; evaporation and, 25, 85–86; exoplanets and, 135–36, 151–52; frost line and, 173–74, 177; habitability and, 85, 135–36; hydrothermal vents and, 5, 15, 29, 32, 39, 42–45, 49, 51–64, 71–73; ice, 2 (*see also* ice); as limited resource, 222; liquid, 15, 25, 38–39, 66–67, 85–86, 135–36, 165, 171, 176–78,

187; Mars and, 25, 85–86, 98, 174; meteorites and, 156, 165–66, 171–72; microcosmos in, 10, 13–15; minerals in, 50, 98, 176–77; *Nautilus* and, 36, 39, 45–49, 100, 115; Noachian period and, 25; oceans and, 37 (*see also* oceans); *Orpheus* and, 70–71; photoautotrophs and, 56; stromatolites and, 83; Underwater Networked Experiments and, 45; van Leeuwenhoek and, 13–14
Wegener, Alfred, 40
Widmanstätten pattern, 160
Wild Dolphin Project, 212
Wirsen, Carl, 57–58
Wollaston, William, 118
Wolszczan, Aleksander, 128
Woods Hole Oceanographic Institution, 70–71

Yang, Stephenson, 125–26
Yellowknife, 175
Yellowstone National Park, 103

zinc, 50, 72
zircon, 175–76
zombie planets, 127